Intellectual
ANARCHY

Intellectual
ANARCHY

THE ART OF DISRUPTIVE INNOVATION

PATRICK K. SULLIVAN

OCEANIT

Oceanit
828 Fort Street Mall,
Honolulu, HI 96813, USA

www.oceanit.com

Intellectual Anarchy™

ISBN: 978-1-7343316-1-5 (Paperback Edition)
ISBN: 978-1-7343316-0-8 (Hardcover Edition)

Ordering Information
Quantity sales: Special discounts are available on quantity purchases by corporations, associations, and others. For details contact the publisher at the address above.

Orders by U.S. trade bookstores: Please contact Ingram Customer Service: Tel: (615) 793-5000; or visit
https://www.ingramcontent.com/publishers/distribution/wholesale

Printed in the United States of America

Publisher's Cataloging-in-Publication data
Sullivan, Patrick K.
INTELLECTUAL ANARCHY: The Art of Disruptive Innovation. [to come] pp. Illustrations.

Subjects: Technology—Science—Innovation—Disruptive Innovation—Engineering—Transdisciplinary Science—Intellectual Anarchy™,—Design Thinking—Human-Centered Design—Mind-to-Market

First Edition

This book is dedicated to the love of my life, Jan Naoe Sullivan,
an amazing, beautiful and smart woman,
my most trusted adviser.

Change is the only constant.
Methods and materials keep evolving;
what was once considered impossible,
may become possible.

CONTENTS

PART III
INNOVATE

PART IV
DELIVER

PART V
DISRUPT

PREFACE

A pessimist sees the difficulty in every opportunity.
An optimist sees the opportunity in every difficulty.
—Winston Churchill

I've been accused of being an optimist.

I believe that the accelerating challenges facing our planet—such as climate change, food scarcity, sustainable energy, overpopulation, nuclear proliferation, income inequality, and rising healthcare costs—are solvable. The question is not if, but when.

So how do we move the future from the distant horizon to the present day?

All these challenges directly impact our lives and our unwritten future. They have enormous economic consequences for humans and society. They also provide incredible opportunities for investors and innovators. But not just "regular" innovators. To exploit these opportunities, we need fearless innovators who are comfortable with ambiguity and risk, who embrace exploration and discovery, who are excited by transdisciplinary thinking, diversity, and by what fields outside their comfort zone have to offer.

We need Intellectual Anarchy™.

"Intellectual Anarchy" is my term for the discipline we developed at Oceanit to persistently deliver disruptive innovation. Intellectual Anarchy provides an essential framework for thinking about what was once unthinkable, a dependable framework to develop and deliver disruptive innovation. The foundation of Intellectual Anarchy is the rule that "there are no rules, only moral and legal guidelines," underscored by the axiom that "the only constant is change."

Intellectual Anarchy is deeply rooted in the scientific method, the most powerful tool that humans have ever created, credited to Francis Bacon (1561–1626), the father of empiricism, but significantly influenced by Copernicus and Galileo. While science is imperfect, subject to the limitations of the humans who practice it and assaults from those who don't understand it, it provides a proven method for creating new knowledge and building the future.

The evolution of our Intellectual Anarchy methodology has been critically informed by our location in Hawaii, far from the groupthink of Silicon Valley. So, how did this all come about, and why Hawaii? As with so many things, the answer is as simple as it is eloquent. There was a girl, beautiful and smart. We fell in love, and after I graduated from the University of Colorado at Boulder, I followed her to her home in Hawaii. We got married and have been happy ever after. I received my PhD in engineering from the University of Hawaii, but then I needed a way to pay the bills.

I liked solving problems, so I launched Oceanit as a consulting company. Keep in mind that this was before Al Gore "invented" the Internet. One might think that this was a doomed situation. On the contrary—it was a great opportunity. It turned out that the only jobs we could initially get involved solving problems that were too risky, too scary, or too complex for other organizations to tackle. This, unknowingly, became the DNA of Oceanit.

Today Oceanit works across the world with just about everybody—the government, Fortune 100 companies, and more than sixty universities—partnering and collaborating as we research, invent, and produce technology in energy, aerospace, life sciences, information technology, sustainability, and resiliency.

During the last quarter of every year, we ask ourselves what issues are interesting, important, and worthy from the position of how we want to spend our time on the planet and how we can impact the world. We refer to this category as "Left of Boom," since we are far to the left of where most people focus—the "boom" of rapid growth with the best internal rate of return or specific requirements from an industry. We approach these issues in the framework of Intellectual Anarchy, bringing to bear new combinations of fundamental science, and we drive to deliver impact to humans and society.

This Mind-to-Market mindset includes building stand-alone manufacturing plants, private equity–financed spinouts, corporate codevelopment partnerships, and other methods of distribution.

Finally, we are prepared not only to innovate technology but also to innovate business models, delivery models, and financing models, deploying techniques such as Design Thinking to tease out the essence of what we are actually offering, and searching to understand how we truly respond to needs that impact the world.

I feel extremely fortunate to have witnessed this process unfold. Moreover, I've been fortunate to do this in a place I love, with people I respect and admire, undertaking projects that are profoundly interesting, challenging, meaningful, and impactful.

/////////

My hope is that this book will encourage more people from around the world to harness science and imagination to capture the future we all need and deserve.

This is why I'm an optimist.

PART I
Question

Innovation does not live in a vacuum, it serves to address need. Solving the *right* problems is critical. Everything begins with asking the right questions.

Some men see things as they are, and ask why.
I dream of things that never were, and ask why not.
—George Bernard Shaw, adapted by Robert Kennedy

INTRODUCTION

This is a book about finding disruptive solutions to hard problems. By "disruptive," I mean solutions that seemingly come out of left field, defy conventional wisdom, and overturn existing industries, practices, and business models. And by "hard," I mean problems so difficult that they can seem impossible. Problems of that magnitude demand radical new ideas—an incremental approach won't work. And they can only be solved by people, however bright or well educated, who are brave enough to embrace risk and operate outside their comfort zone.

I always ask new scientists and engineers joining our organization what they want to accomplish with their time on the planet. Often, they're stumped. They think it's a trick question, but it's very serious. We all have a limited time here on Earth—we have not solved that one . . . yet. So, if we're going to make a difference, make a positive impact on humans and society, we need to make some hard choices. If we're going to solve climate change, cure disease, achieve routine space travel, or create an artificial intelligence capable of more than playing Jeopardy or Go, we need to focus. We need to execute a disciplined approach that produces results in our lifetime.

We call the approach we've developed at Oceanit—after thirty years of tackling these seemingly intractable problems and coming up with creative solutions—Intellectual Anarchy. It's as much a mindset as a discipline that enables persistent delivery of disruptive solutions to the world's hardest problems.

After practicing Intellectual Anarchy for so many years, I'm able to sort people into two categories: those who are terrified by the challenges ahead and those who are energized. For those who are terrified, I make no judgment. For those who are energized, I've found an approach that unlocks their potential to be extraordinary, producing a new breed of innovators that I call Techno Warriors.

//////////

Curing cancer is the quintessential hard problem. As a society, we've invested more than half a trillion dollars since we declared war on cancer in 1971, with no end in sight. Mortality rates have fallen since the 1990s, but the reduction has been less than 2 percent per year. We still have a long way to go.

Part of the problem is that what we call cancer is really a constellation of related diseases, all similar in mechanism but distinct enough that they generally have to be tackled individually. What works for melanoma won't work for a brain tumor; what works for leukemia doesn't work on breast cancer. As a result, oncologists have developed an arsenal of techniques for attacking cancer by various means: surgery, chemotherapy, radiation, diet, tissue replacement, immunology, and so on. Each of these has chipped away at one form of cancer or another, increasing survival rates, but none amount to a cure. They are incremental improvements only. What we need is a disruptive solution.

When the cure for cancer arrives, it's likely to come from somewhere unexpected, outside the medical field entirely, such as artificial intelligence, nanotechnology, or robotics. This is one of the hallmarks of disruptive innovation. Steady, incremental progress comes from within a field or industry, but wild, seemingly unpredictable, disruptive innovation comes from elsewhere. This is a major motivation for transdisciplinary thinking, one of the pillars of Intellectual Anarchy that I discuss throughout this book.

An example from science fiction of what such a disruptive solution *could* look like—we're not there yet—is in an episode of *Star Trek: The Next Generation,* which deploys transporter technology in clever fashion to cure an "incurable" disease. They selectively edit the patient's beam pattern with a previously stored beam pattern from before they contracted the disease. The cure comes not from the sophisticated medical advances of Star Fleet but from their transporter technology. Another example is in the 1966 film *The Fantastic Voyage*, where miniaturization technology developed for Cold War espionage is used to put a manned submarine into a patient's body to attack an inoperable blood clot from inside the brain itself. In both cases, the medical benefits are incidental to the intended use.

At Oceanit, we're betting on artificial intelligence or nanotechnology beating medical science to the cancer cure. In a nod to *The Fantastic Voyage,* we've created an autonomous "submarine," a nanodevice capable of traveling through a patient's bloodstream to deliver a precisely calibrated dose of chemotherapy drugs directly to a cancer site. Chemotherapy works by exploiting cancerous cells' increased vascular demand, essentially poisoning the patient and hoping that the cancer cells will take up more of the toxins than the healthy cells and die first. But it's imprecise, and the side effects range from debilitating to devastating. By delivering chemo directly to the cancer cells with a "lock & key," the macro dosage can be reduced by a factor of 100,000 because the delivery specificity has been increased 100,000 times. This precise delivery reduces "bystander effects," like losing your hair and getting sick, while maintaining efficacy against cancer cells. We call this NANOBE, and it is available today, being used in laboratory trials and experiments. Another approach we're actively pursuing involves the use of language-based artificial intelligence to divine the "intention" of cells and intervene before they become cancerous. The best cure for cancer is not getting it in the first place.

None of this is meant to diminish the brilliant work that cancer researchers are doing in their field nor the progress they have made; it is only to suggest that disruptive solutions require unconventional tactics that involve going outside the lines.

Hard problems can be found in every industry, from health to energy to transportation and more: curing cancer, reducing greenhouse gas emissions, preventing oil-well blowouts, and so on. This book details the Intellectual Anarchy methodology that we follow at Oceanit for generating disruptive solutions to these problems and taking them from the idea stage all the way to market—a process we refer to as Mind-to-Market.

No mere academic posturing or armchair philosophizing, Intellectual Anarchy is a tried-and-true discipline that has evolved from over thirty years of solving hard problems for ourselves and our customers, resulting in breakthrough products and technologies such as LATCH, a laser-based system for halting brain bleeds without surgery, or HeatX, a surface treatment that can significantly improve the efficiency of a heat exchanger.

Much of this methodology is informed by our location in Hawaii, far from the groupthink of Silicon Valley. Hawaii is a chain of islands in the middle of the Pacific, about 2,500 miles from the West Coast of the US mainland. Whenever I fly home, I'm reminded of its physical isolation. But modern conveniences like fast, cheap air travel and broadband Internet keep Hawaii connected to the rest of the planet and have largely eliminated the tyranny of distance. And what Hawaii has to offer is immense: a rich and diverse culture, an incredible environment, and the University of Hawaii, an excellent research university.

In this postmodern world, an island is as much a metaphor as a physical place, and what we have learned in Hawaii can be applied to most of the world. Every major US state has at least one great research university, and there are nearly 5,000 colleges and universities across the United States offering STEM education. Most of the US population, around 90 percent, lives far outside the traditional tech hubs of Silicon Valley, New York, and Boston. They are outsiders, too—islands in the metaphorical sense, as isolated as Hawaii. And they can find value in our approach and the discipline of Intellectual Anarchy.

//////////

Innovation, particularly disruptive innovation, is not something businesses necessarily want—but it is something they need. When Alfred Sloan articulated what has since become the core of the modern MBA, the world was different. Management strategy at large conglomerates focused more on reducing risk and increasing predictable outcomes than innovation. This strategy was sufficient for an early industrial world concerned primarily with mass production and efficient distribution, and it enabled Sears, for example, to become the superstore of America. But in our increasingly connected global economy, it falls short.

Remnants of the past industrial world are still on display in places like Ohio, Pennsylvania, Michigan, West Virginia, and Wisconsin, the so-called rust belt. Large businesses that dominated their markets believed they had no need to change. And who could blame them? Change slows you down and adds risk. They were accustomed to the status quo. But that proved to be a shortsighted view.

I was recently in Buenos Aires, where I took a bike tour around the city. Today it's hard to believe that Buenos Aires was once a very wealthy city—one of the wealthiest in the world at the turn of the last century. Vestiges of this incredible wealth can still be seen everywhere, but the world changed, and wealth shifted across the planet. It's the same story with different details for cities like Riverside, California, the birthplace of the navel orange, which back in 1885 was considered the richest city per capita in the United States, until the world changed again; or Detroit, Michigan, the "Motor City," once the center of innovation and production for the automobile industry.

The specific stories are different, but the basic issue is the same. Change happened. It's the only constant. And it's happening faster and faster. As a result, what I call "the kinetics of innovation"—the complex nature of ideas, market access, connectivity, and speed—are recasting the future. The only question is how will you ride the change and capture the future: Will you keep pace or fall behind and be swept aside?

This book shares the hard-won lessons I've learned over the past thirty-plus years of building a successful, bootstrapped technology company that doesn't just survive innovation, it thrives on it.

Who Should Read This Book

The pace of innovation is accelerating, and those who can't keep up will be left behind. This book is not just for innovators; it is for anyone concerned about the future and the potential of breakthrough innovation. You should read this book if you are

- a person who wants to impact humans and society / save the planet;
- a community that is worried about out-migration of their best and brightest;
- a mature enterprise that has lost, or is at risk of losing, its competitive edge;
- in need of creative solutions to difficult problems;
- looking to break free of the groupthink of Silicon Valley;
- located outside the traditional tech hubs and frustrated by the lack of access to resources.

How This Book Is Structured

Each chapter of *Intellectual Anarchy* is divided in three parts. The first is an explanation of one or more of the basic principles of Intellectual Anarchy. The second is a case study of a particular Oceanit project that serves as an exploration of that principle in action. The third spells out the implications of the specific innovation and how they could inform the future. The final chapter, "Left of Boom," serves as an integration of all the principles and demonstrates how we apply them in order.

PART II
Think

The human brain is the most powerful device in the world. Disruptive innovation requires becoming comfortable with how to use the human brain to think. This is how one small company on an island in the middle of the Pacific thinks. This is Intellectual Anarchy.

> *I think, therefore I am.*
> —Descartes (1641)

CHAPTER 1

How We Think:
No Rules, Just Moral and Legal Guidelines

You don't learn to walk by following rules.
You learn by doing, and by falling over.
—Richard Branson

Anyone who has never made a mistake
has never tried anything new.
—Albert Einstein

At the dawn of the twentieth century, the world was racing to invent the airplane. Aeronautical experts and engineers from around the globe competed to develop heavier-than-air flight. First across the finish line, though, was a pair of humble bicycle mechanics from Dayton, Ohio.

The Wright brothers made the first manned, powered flight at Kitty Hawk, North Carolina, in 1903. It lasted just twelve seconds, but it changed the world. Commercial passenger and cargo flights began almost immediately. When World War I broke out in 1914, airplanes were pressed into use for reconnaissance, then quickly evolved into bombers and fighters as countries vied for air superiority. The first transatlantic flight took place less than a year after the cessation of hostilities, shaving travel times from days to hours and paving the way for ever-increasing globalization.

Disruptive Innovation

Clayton Christensen first coined the term "disruptive in-novation" nearly twenty years ago in his landmark book, *The Innovator's Dilemma*, to explain, very narrowly, the destruc-tion of existing markets by low-end entrants, but it has come to mean any sweeping innovation that disrupts an industry. Throughout this book, I use the term in this broader, modern sense.

Disruptive innovation pushes science and invention past the bounds of what's considered possible. Disruption exists on the threshold between failure and discovery. It's not easy, but with practice, one can learn to get comfortable on the edge of the knowledge abyss.

The airplane was a disruptive innovation, an advance not just of degree but of kind. It created a profound shift in the status quo: accelerating travel and shipping, dominating military strategy, and shrinking the globe—and transforming how humans and society interact.

Another example, closer to home for most of us, is the iPhone. Apple didn't just improve on existing phone designs, it revolution-ized personal communications and captured nearly 95 percent of smartphone industry profits in the process. It similarly transformed the way humans and society interact. The consequences of that disruption are still being worked out.

Many experts and industry leaders predicted the iPhone would fail. The iPhone, with no physical keys, didn't fit their preconcep-tions of what a phone should be; Apple, with no prior experience in the industry, didn't fit their preconception of what a phone company should be. But Apple and the iPhone succeeded *because of* these issues, not in spite of them. Today the iPhone is the most successful consumer electronics product of all time. Every phone looks like the iPhone and every consumer electronics company is scrambling to become the next Apple.

We've become accustomed to the airplane and the smartphone. They no longer feel disruptive. But now imagine the following:

- an injectable machine the size of a blood cell that fights cancer;
- a road made of intelligent concrete that knows the weight and speed of vehicles that travel on it;
- a laser-based system that can perform surgery without breaking the skin.

These are all real projects in various stages of development at Oceanit, each of which has the potential to transform the world.

These projects were all developed via *Intellectual Anarchy*—a counterintuitive strategy that lends itself to persistent, disruptive innovation: *Intellectual* because it's rooted in deep science and human-centered design; *Anarchy* because it "breaks the rules" and defies conventional wisdom. Developments such as the Wright Flyer and the iPhone also display elements of this method. The important thing is that Intellectual Anarchy is a discipline, a methodology that can be practiced, leading to consistent, predictable results—versus serendipitous, one-off, eureka moments.

I explain exactly what I mean by *deep science* and *human-centered design* later on in the book, but right now I want to introduce one key idea that we'll return to again and again: favoring a *transdisciplinary* approach over the traditional expert approach. This is a core principle of Intellectual Anarchy that I believe fueled the success of the Wright brothers and Apple, and that I know was at the heart of our many breakthroughs at Oceanit.

> *When I let go of what I am, I become what I might be.*
> —Lao Tzu

Experts, by definition, are good at what they know, but may have little to offer on what they do not know. In the transdisciplinary approach, we separate expert skills from expert knowledge. We deliberately introduce the experts into an environment where their expert knowledge does not apply, enabling them to take a fresh look at the problem, while still bringing their expert skills to bear.

Our Western educational model focuses students on acquiring preexisting knowledge rather than inventing and discovering for themselves. Students are discouraged from doing original research until they reach the PhD level, and even then they are discouraged from straying too far into the unknown. But the transition from the known to the unknown is precisely what's important. How to cross that threshold is what needs to be understood and replicated, not merely registered as a milestone.

In academia, once this threshold is crossed, students/researchers tend to stay in their particular lane. Transdisciplinary thinking encourages them to move to another lane and repeat the breakthrough thinking process. This holds not only for PhDs but for anyone who needs to solve a big problem.

Applying problem-solving skills to a new field can be uncomfortable for experts, deprived of what they think of as their greatest asset—their expert knowledge. It requires confidence, and it requires understanding that the problem-solving skills they've acquired are as valuable, if not more valuable, than the knowledge they acquired developing those skills.

Wright Brothers Carved Their Own Path

The Wright brothers defied much of the conventional aeronautical wisdom of the time. While their competitors relied on data passed on from other aviation pioneers (for example, the Smeaton coefficient of air pressure in the lift equation), the Wright brothers found it didn't agree with their own informal observations. They built their own wind tunnel to facilitate direct observations on scale models and calculated the Smeaton coefficient correctly for the first time.

Their competitors also envisioned airplanes that sailed through the air like ships controlled by a rudder, which led them to design wings that would reinforce stability in flight.

The Wright brothers took their cues from birds, rather than existing designs, and relied on the pilot to stabilize their craft in flight. They created the first three-axis control system (for which they received a patent) and invented the idea of variable-geometry wings, both radical innovations for the time.

Although the Wright brothers consulted with experts by mail and occasional in-person visits, I believe their geographical isolation from the aeronautical hotspots contributed to their independent thinking—an idea revisited in chapter 3 on groupthink and chapter 6 on the geography of disruption.

The Wright brothers weren't trained aeronautical engineers. But their engineering and mechanical skills transferred easily to airplanes. While they didn't outright reject the conventional thinking on how an airplane should be designed, neither did they blindly follow the lead of their predecessors and other experts. Their willingness to question what others took for granted, to observe and experiment, and to carve their own path ultimately led to their success.

Apple had no previous experience with phones. When the iPhone was announced, Palm CEO Ed Colligan famously said, "PC guys are not just going to just figure this out. They're not going to just walk in." But Apple's legendary design skills plus their fresh perspective on what a smartphone *could be* made the iPhone a home run.

At Oceanit, we have science and engineering experts in a wide variety of fields: astrophysics, electrochemistry, civil engineering, artificial intelligence, marine biology, and more. Whatever the problem domain, we have the expertise to tackle it. But whether we assign those experts to the problem at hand depends on the level of the problem.

We divide problems into three categories: interesting, challenging, and disruptive. Interesting problems are ones that any competent engineer or scientist could handle; for example, a simple productivity app like a to-do list or a calendar. We choose to pursue these on occasion because they keep us grounded with firm deadlines and concrete deliverables. Challenging problems are complex due to logistical, geographical, or supply chain issues and require a higher level of management and understanding; for example, an international e-commerce solution that provides secure transactions over multiple languages, currencies, and international banking regulations. And finally, there are disruptive problems—high-risk, swing-for-the-fences, put-a-dent-in-the-universe projects. They carry no guarantee of success but are worthwhile because if they pan out, the benefits will be enormous and transformative.

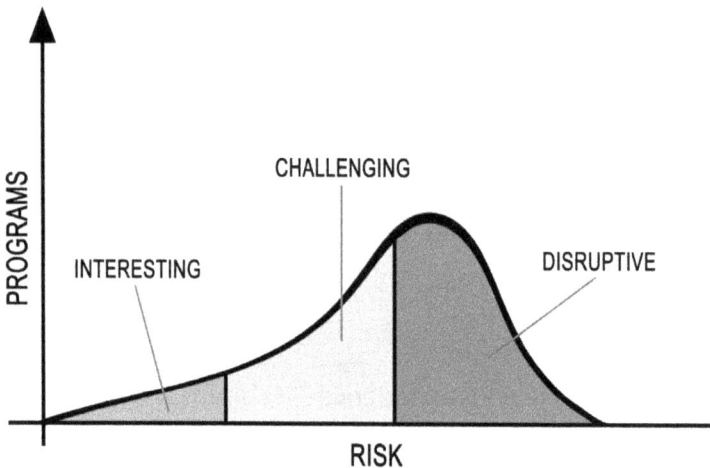

Fig. 1.1. We divide programs into three buckets. Interesting programs are those that nearly any competent engineering or technology organization can execute. Challenging programs are those that are both interesting but also complex in execution that might be logistically challenged, and require high-quality management teams. Disruptive programs are those that push the edge of science—typically, high-risk ideas that, if successful, move the needle and create new insights that reveal the future.

For either of the first two categories, we want our experts on the team, as would most organizations. But when a disruptive solution is called for, we do not limit the team to subject-matter experts. Sometimes we deliberately *remove them* from the initial team formation.

For instance, we recently launched a prosthetic brain project, intended to assist patients with brain damage analogous to the way that a prosthetic limb assists an amputee—by providing missing functionality. For example, a patient with face blindness, unable to recognize the faces of family or friends due to damage to a particular part of the brain, would use a prosthetic brain to handle the facial recognition for them. As a first step, we developed a neurological sensor to pick up brain activity to see if we could control external devices directly from the patient's thoughts. Initially, we brought in our own in-house neurological expert— who promptly informed us that what we were attempting to do was impossible. So we assigned the expert to another project and pressed on without him. It's hard to motivate individuals to work on a project when they believe it can't succeed or a team to take on a challenge when the expert tells them it's impossible. But only by engaging with this "impossible" problem could we learn to surmount it. We needed to try, and fail, to learn through failure and try again, unhampered by expert doubts. Eventually, we discovered a completely new method and succeeded in building the device that "couldn't be built."

We then reintroduced our neurological expert to the project so we could translate our findings into the more traditional language familiar to the medical field.

While our team found the presence of a *neurological* expert confounding, they were deeply experienced in applied physics, electrical engineering, and signal processing. Transdisciplinary thinking is not a rejection of expertise per se. It is an end run around the preconceptions and narrow focus that can accompany expertise. Because we were attempting disruptive innovation on a neurological problem, it was our neurological expert that had the biggest blind spot, believing he already knew what was and wasn't possible to achieve. Our experts in other disciplines were free to lean on their mastery of their own areas and comfortable taking risks and exploring new ideas in a field that wasn't their own. They *didn't* know what couldn't be done in neurology, and so they were able to bring to bear their own expertise in attempting to find a solution.

This is how we think at Oceanit.

Intellectual Anarchy is a mindset, an approach to problem solving we developed from fundamental concepts, that has worked consistently for us at Oceanit for thirty years. Of course, there's more to it than just throwing nonexperts at a problem. Intellectual Anarchy requires highly skilled yet flexible thinkers and a supportive environment that rewards risk and encourages transdisciplinary thinking over expert thinking. I discuss all of these factors in more detail in the chapters ahead.

Magic Dirt: Rapid Control of Severe Bleeding

Few technologies connect military and civilian populations like medicine. The extreme conditions of the battlefield drive innovations that ultimately make their way into civilian society, resulting in cheaper, more effective treatment—increasing life expectancy and improving quality of life. One example is hemostatic agents (from the Ancient Greek *haima*, meaning "blood," and *stasis,* meaning "still"), materials designed to halt rapid bleeding, which date back to the Trojan War in the thirteenth century BC.

Blood loss from trauma is the most common cause of death among American combat troops. The logistical difficulty of treating casualties in a war zone means that sometimes soldiers die of potentially nonfatal injuries, resulting in unnecessary deaths from wounds that could be treated if the right treatment were available. Over the past decade, nearly 25 percent of American forces killed in action died from wounds they could have survived if treated in time. This is a particular problem with forward-deployed Special Forces who may find themselves behind enemy lines with no recourse to medical attention. Finding better ways to quickly control bleeding on the battlefield is a high priority issue for the US military.

Battlefield Casualty Rates

The Iraq War drew criticism for (among other things) its horrific injury rate from roadside bombs. Americans were shocked by images of returning soldiers with multiple missing limbs. But these images, traumatic as they were, masked a deeper truth: the battlefield had actually become *safer* for American soldiers due to improvements in battlefield medicine.

The battlefield has always been a dangerous place. Before the invention of antibiotics, even a minor wound, treated immediately, could lead to an agonizing death days later from infection. Amputations to halt the spread of gangrene were common—and not always successful.

WAR	Killed in Action	Wounded in Action	Death Rate
WWII	291,557	670,846	30%
Korean War	33,686	92,134	27%
Vietnam War	47,424	155,303	23%
Iraq War	3,836	32,222	11%

During World War II, nearly one-third of American soldiers that sustained injuries died as a result. That number has crept down over the years, but improved dramatically for the Iraq War, largely due to more rapid and successful treatment of critical injuries. Soldiers surviving with missing limbs, while severely disturbing, is an improvement over wars past, when they likely would have died.

Currently, medics are equipped with hemostatic agents that speed the coagulation process and serve as a last-ditch defense against severe bleeding. In cases where a tourniquet is inappropriate and direct pressure is insufficient, the application of a hemostatic agent can mean the difference between life and death.

Until fairly recently, the hemostatic agent most in use by the military was a loose, mineral-based powder. The active ingredient, zeolite, selectively absorbed water from the blood, thickening it and thereby increasing the density of clotting agents. It had the unfortunate side effect of producing an exothermic (heat-generating) reaction that could cause second-degree burns. The powder also had a tendency to adhere to the flesh, posing a challenge in debriding the wound without additional damage to already injured tissue.

Within the past few years, the military has largely switched over from zeolite-based powder to a hemostatic dressing or combat gauze—a bandage treated with an additional agent to quickly stop bleeding. The dressing most commonly in use relies on kaolin, a claylike mineral, which activates natural clotting factors in the blood, speeding coagulation. Combat gauze is easier to use correctly and doesn't entail the risk of chemical burns, though it does pose the problem of nonbiodegradable kaolin leeching into the blood from the dressing.

The bigger problem with both of these products is that they are stopgap measures, designed to stabilize a patient just long enough to get them to proper medical care. What the military really wants is a total solution—a bandage that not only stops bleeding but also disinfects the wound, prevents infection, alleviates pain, and accelerates the healing process—allowing the soldier to stay in the field.

Approach to Hemostasis and Caveats

The ideal hemostatic agent would be easy to use,
highly efficacious, fully absorbable, and inexpensive.
Unfortunately this agent does not exist.
—Achneck, et al., *The Annals of Surgery*

In the first century AD, a Greek physician with the unlikely name of Rufus of Ephesus listed all of the methods then known for controlling hemorrhage as "compression, styptics, the cautery, torsion, and the ligature." In other words, tying, kinking, burning, direct pressure—and hemostatic agents. All these techniques are still in use today. Of these, hemostatic agents have advanced the furthest in the last 2,000 years, with much of that progress taking place in the last two decades.

As miraculous as they can seem, current hemostatic agents are a treatment of last resort, intended to be used when conventional methods are insufficient, and then in *addition* rather than *instead of*. Even the best hemostatic agents available today are less effective than direct pressure and tourniquets.

We learned of this issue in various conversations with military personnel at all levels, from boots-on-the-ground soldiers through program managers. We particularly like to talk to soldiers at the tip of the spear because they're the ones living with the day-to-day reality of decisions made higher up. They don't complain, but they have an unflinching view on what really works. We listen to them carefully and get to know how they're doing.

We already had a team from Oceanit working with the US Army on another project. We had just demonstrated a successful prototype of a nanochip for detecting potential adverse reactions to an anthrax vaccine. We asked our team to investigate the possibilities for improved treatment of severe wounds.

Each of our teams is led by a champion, someone who passionately believes in the project and will lead the charge despite the obvious challenges ahead. As discussed in chapter 1, this champion is often someone working outside his immediate area of expertise. Too often, we find, experts will preemptively limit the potential to "plausible" avenues of investigation. In seeking out disruptive innovation, we sometimes, like Sherlock Holmes, have to consider the implausible.

In this case our team champion was Mary Z. She'd earned her PhD from the John A. Burns School of Medicine right here in Hawaii, but her expertise lay in pathology and toxicology, not hemostasis.

Mary had great lab skills and had been an excellent supporting member of several successful teams at Oceanit in the past and had led some projects involving neurotoxins.

Mary's teammates were an eclectic mix of scientists and engineers with backgrounds in nanomaterials, physical chemistry, neuroscience, bioengineering, and high-energy physics. As usual, we tried to assemble a diverse team, pulling members from many different disciplines, because we find greater success with input from many varied perspectives.

Mary and her team started with the idea of an all-purpose wound-control system—the "total solution" envisioned by the military. They began development on a gel laced with antibiotics, analgesics, steroids, and a proprietary hemostatic agent. The gel was designed to be poured directly into the wound. They also explored the idea of embedding the gel in a matrix, which could then be applied like a bandage, similar to combat gauze. We called this superbandage *SwiftSkin* because it could heal a wound 200 percent faster than typical bandages in early trials.

This played into a natural strength of Oceanit. We've done a lot of work with functional fabrics through our custom, nanomanufacturing process. Many of these projects began as simple experiments, looking at interesting ideas we could explore, whether or not they had any intrinsic value—like creating a fabric that could produce energy. Others have led to successful products, like DrySAF, which can absorb up to ten times its weight in water and allows US Marines to dry their standard issue RAT (Rugged All Terrain) boots in a fraction of the normal time.

But this time, the synthesized functional fabric didn't work as intended. It had a tendency to crumble and fall apart—not exactly what you want in a field dressing. The team hit on the happy idea of grinding up the bandages and trying it anyway. The resulting material actually resembled freeze-dried coffee or a pile of fine dirt.

The team tested this "dirt" on pig livers picked up at a local butcher shop in Chinatown. They circulated blood through a liver, then simulated an injury with a deliberate incision. Packing the wound with "dirt" stopped the bleeding with astonishing speed, taking less than 5 percent of the time of the best combat bandages available. In fact, it worked so well that early graphs of the results looked like errors. Even better, the "dirt" was completely biodegradable, meaning it could be left in the wound where it would eventually dissolve on its own.

We commissioned a study to perform an independent check on our findings. A researcher at the John A. Burns School of Medicine performed similar tests on mice and achieved the same results, confirming the efficacy of our magic dirt at halting blood loss.

Playing on the appearance of the compound, we called it DERT, which stands for *Dispersible Enhanced Recovery Treatment.* Oceanit is currently in discussion with the US military about making DERT available to combat troops, as well as with a major pharmaceutical company about potential civilian applications—for ambulance crews, home emergency kits, even operating rooms, where there is a risk of severe bleeding during surgical procedures involving the kidneys, spleen, liver, and so on. Currently, these surgeries require a thrombin-based hemostasis. Thrombin is an expensive enzyme produced from animals. We can produce DERT for peanuts, and it works nearly as well as the best thrombin-based product available today, currently reserved for special surgical applications because of its high cost.

We did run into one problem with DERT, however, a problem common to innovators. The compound is so unique that it can't be manufactured by conventional methods. We had to create a new way to manufacture and distribute it. But that's a story for another chapter.

//////////

Implications

The implications of DERT are a microcosm of the technological, economic, and regulatory issues that keep truly affordable health care out of our reach. Remember, DERT was originally developed for wound management in forward-deployed Special Forces troops beyond the reach of hospitals and at risk for injuries that are difficult to treat. Whether from a traumatic bleeding gunshot wound or minor nick to the liver or spleen, death from blood loss occurs within minutes and can't be treated with conventional bandages. Packing wounds with hemostatic DERT can stabilize patients until they can be transported for surgery. However, it's also designed to stay in the wound and dissolve over time, supporting the healing process.

One of the key advantages of DERT is its low production cost, which translates into availability and affordability for users. That's not how things work in the civilian health care industry, however. High-end hemostasis agents that typically contain thrombin are incredibly expensive. As a result, their use is reserved for the most acute patients, fully insured patients, or full-pay patients. Our early work showed that DERT performs similarly but costs pennies to produce.

Nevertheless, DERT is still unavailable to the surgical suite today. There are various reasons, but the lack of capital to get it there tops the list. As a "rule of thumb," bringing a new medical device to market requires significant capital—roughly $50 million and eight years for a medical device and $150 million and ten years for a therapeutic treatment (provided that the process is executed flawlessly), making the path to market prohibitive for most. As of the middle of this decade, the median cost of trials for a new drug to meet FDA approval was $19 million, according to the journal *JAMA Internal Medicine*.

These challenges tend to protect the current franchise holders, creating a monopolistic position for current "big medical, big pharma" that drives health care prices upward. It's good for those businesses, but not so good for the rest. You can see blatant abuse of this franchise position in companies such as Turing Pharmaceuticals, which produces Daraprim, used to treat AIDS patients. Turing increased the price of Daraprim from $13.50 to $750 per dose in 2015. Later its CEO, Martin Shkreli, was convicted and sentenced. However, his conviction was more about securities fraud and other abuses rather than price gouging, which is perfectly legal; and the price per dose remains at $750 today. This behavior, which Gordon Tullock and Robert D. Tollison refer to as "rent seeking," inhibits innovation by prioritizing profit over the idea or innovation itself.

US health care costs continue to escalate, closing in on 20 percent of the GDP and an annual cost of nearly $9,500 per capita, making it the most expensive in the world, yet delivering inferior results in many areas. We rank thirty-third out of thirty-six countries with recorded OECD data (after New Zealand and the Slovak Republic) in infant mortality and twenty-eighth (after Poland and Turkey) in life expectancy (2018 OECD data).

The implications of disruptive innovation for making health care more effective and affordable are enormous. Therefore, the debate about health care delivery, fueled by those franchise holders protecting their business, is also about the business process of bringing disruptive and revolutionary medical technology to market. There is virtually no incentive for a health care company that provides products and technology for the surgical suite to reduce its costs. Although it's typically defended as "top of class" care, it's got everything to do with business—and money.

Civilian health care has traditionally been much slower to innovate than military medicine. As health care expands to consume an ever-greater percentage of the GDP, disruptive innovation becomes more important in keeping costs in check.

Health care is ripe for a major revolution, something that is highly customized and highly specific for individual patients, but inexpensive—priced like a commodity. It will be driven by disruptive technology, similar to the personal computer revolution that put a powerful computer in everybody's hand. The arc of technology will break through the barriers protected by the current franchise holders.

As one example of what I mean, former DARPA Biological Systems program manager Dr. Geoff Ling is working on a drugstore vending machine for certain medicines that would compound and dispense medications on demand. This would cut costs dramatically and improve availability. Essentially, he is developing the equivalent of a 3-D printer for prescription drugs that could ensure the correct medication is available at the time that it is needed.

Business models and incentives will change, too, killing entire categories (e.g., shifting from treating chronic disease, which locks in a guaranteed revenue stream, to curing it outright). However, this revolution is likely to come from outside traditional health care companies, just as the iPhone came from a company with no phone experience.

Here are two mechanisms I can see for bringing about this revolution. Implications are pointing to several opportunities, including the following:

1. Launch a health care–focused version of DARPA—HARPA—to actively underwrite disruptive, high-risk, out-of-the-box experimentation, discovery, and development with potential high payoff health care impacts, resulting in more affordable and effective health care.
2. Create a government-backed finance mechanism with shared economics to bring promising medical tech to market. This could range from Private-Public-Partnerships (P3) to loan guarantees. For example, it could include FDA approval and a lock-up provision to discourage flipping technology to big medical acquirers. This could be modeled after the US Department of Defense program enabling the manufacture of critically needed material and technology, as authorized by the Title III Defense Production Act.

///////////

Whatever your industry, disruption and evolution start with looking at the way things are and wondering why. You might just find that the problem isn't what you thought it was, and the solution—along with your opportunity to provide that solution—is where no one expected to find it; where no one else was even looking.

The implications of "how we think" have an oversized impact on innovation. How we view a partially filled glass of water—whether half full or half empty—can overshadow how we see the world, when really, the important questions might be, "What is in the glass?" or "What else can the glass be used for?"

Creativity and innovation are skills exercised early in life; we are all born with this ability. Children are naturally curious—born to explore and discover. Although we are born natural innovators, we somehow surrender this trait while growing up. Nevertheless, it can be relearned, and this begins with taking ownership of how we think, our willingness to experience the world, and how we perceive risk and failure—how we experience life.

I can summarize with a surfing metaphor: innovation can feel like paddling into a wave that scares and amazes you at the same time: it's truly exhilarating. If you don't make it—wipeout—you hope you don't get injured. However, the excitement of the ride is worth the risk. Disruptive innovation has a similar adrenaline rush.

Those who are willing to pursue disruptive ideas are generally highly intelligent, with talent of equal or higher caliber than you would find at an excellent university. They can navigate and manage complexity. They also have a "risk appetite" for disruption, which exists at the fine line between success and failure—an extremely uncomfortable zone for most. Finally, what drives success in disruption is a very personal decision calculus: "Will I regret not trying?" for such individuals surpasses "What if I fail?" This is because disruptive innovation pushes science and invention past the bounds of what's considered possible, to the line between failure and discovery. It's not easy, but with practice, one can learn to get comfortable on the edge of the knowledge abyss.

How one elects to think ultimately determines who will be a successful innovator. One large study involving entrepreneurship training in the United States found a correlation between risk tolerance that encourages innovation and the benefits of entrepreneurship training. The study found that the risk-tolerant individuals started more new businesses and were more comfortable working autonomously than others. We can conclude from this that true innovators accept high risk for high reward and don't mind going it alone. Going back to my surfing metaphor, that's exactly what it's like: taking on and catching that epic wave.

CHAPTER 2

The Man Behind the Curtain:
The Hazards of Expert Thinking

If an expert tells you it can't be done, get another expert.
—David Ben-Gurion

There's an old joke about a man who loses his keys. A police officer happens on him searching for them under a streetlight and offers to help, asking, "Are you sure you lost them here?" The man says no, he lost them a block over, "but the light's better here."

We laugh at the idea of someone looking for their keys far from where they lost them, but we only have to change one minor detail of the story to make it depressingly relevant: in the hunt for disruptive innovation, we don't know where we lost our keys, or if there even are any keys to be found. Breakthroughs of that magnitude are rare and unpredictable, and we're so afraid of wandering in the dark that we're tempted to confine our search to familiar, well-lit terrain.

That's just one failure mode of a phenomenon I call *expert thinking*, a cluster of cognitive biases that can affect scientists and researchers with expert-level domain knowledge.

I want to be clear that I'm not dismissing experts or expertise per se—only the traps that *some* high-level experts can fall into when pursuing disruptive innovation. When I get a cavity filled by my dentist, I'm not looking for a radical new technique. He may have done 10,000 fillings in his career already, so I know he'll be able to handle mine quickly without pain or complications.

PATRICK K. SULLIVAN

Experts are also essential to innovation. They know the tools and methods of the field, enabling a rapid start. No time is wasted covering well-trod ground. They know the conventions and terminology of the field and are able to communicate with other experts (essential to scaling up, as noted in chapter 13). And experts are familiar with the state of the art. If something can be done, they'll know it, and they'll probably know the most efficient way to achieve it. If you're developing a cutting-edge gene therapy, for example, you're going to need a DNA sequence. You'll save time and energy by employing an expert geneticist to handle the sequencing for you. The sequencing itself may not be the area where you have to innovate, just a necessary step in the process.

But these advantages come at a price.

Because experts are so knowledgeable in their own fields, they are reluctant to stray outside it, prematurely limiting the solution space when you *do* want creativity and disruption. That same geneticist who saved you so much time on sequencing is likely to balk at analyzing DNA differently from the current standard practice: "This isn't how we do it." This can be attributed to natural bias or to the opportunity cost of straying too far from their area of expertise—why fumble in the dark when you can cover so much more ground in the light? But, sadly, this is an attitude inculcated in academia—students are many times actively discouraged from branching outside their discipline—and the habit is hard to break once acquired. Too much time spent in a rigid academic setting, though enlightening, can also be crippling. Finally, experts, by dint of the massive expenditure of time, effort, and money spent in acquiring their education, can have a great deal of their self-worth tied up in their perception of themselves as experts, making them very reluctant to risk feeling foolish by stepping out of their comfort zone. It's a matter of mindset. The education acquired enables standing on the shoulders of giants to see beyond, but that's the beginning of the journey, not the end it is oftentimes perceived to be.

As an example of how perceived opportunity cost can limit more adventurous thinking, look at how long it took the petroleum industry to adopt hydraulic fracturing (fracking). Their vested experience in conventional oil extraction techniques from convenient reservoirs made them reluctant to go after the "inaccessible" oil and gas deposits in shale formations. It took the pioneering efforts of George Mitchell, the "father of shale," to demonstrate its economic viability and the threat of "peak oil" to drive the big producers to his door. Fourteen years later, the United States is a net exporter of petroleum.

These are, of course, tendencies and not absolutes. There are excellent scientists and engineers, experts all, who defy this description. But the tendency is real.

To be fair, if the goal is incremental innovation, a modest evolution of existing science or technology—and most of the time it is—then an expert is what we want. An expert is what we need. But if we want disruptive innovation—revolution, not evolution—then expert thinking can become a straitjacket.

Experts can be too quick to render a particular innovation as "impossible." Because they know their field and are familiar with what's plausible within its parameters, and because they can be so skittish about venturing outside those parameters, if they can't see how something *can* be done they are premature in declaring that it *can't* be done. It is far safer—for their egos, for their reputation—to decide that something is impossible than to admit that it might be possible, but they can't see how to do it. They are, after all, experts. If it could be done, they would know (and, conversely, if they don't know, then are they really experts?)

Heavier-than-air flying machines are impossible.
—attributed to Lord Kelvin

Unfortunately for experts, history is filled with their pronouncements that this or that technology could never succeed, that the end of physics—or of science, or of patentable inventions—was at hand. Look on the Internet and you can find any number of historical false predictions from men of science claiming that powered flight or space travel or nearly any other innovation you care to name is impossible, forever out of the reach of humankind. But keep in mind that as materials and methods change, what was once impossible may become possible. As the kinetics of business speeds up, so does the rate of change of methods and materials. This accelerating change is both exciting and scary, particularly for "experts" who are not constantly tracking, learning, and experimenting with new methods and materials that will likely come from adjacent disciplines and fields of study.

I want to examine one of these assertions in more detail because it illuminates precisely how the expert mind is led astray.

In 1926, A. W. Bickerton, professor of physics and chemistry at Canterbury College in New Zealand, dismissed the idea of space travel, specifically a moon shot, as "foolish" and "basically impossible," saying, "Let us critically examine the proposal. For a projectile entirely to escape the gravitation of earth, it needs a velocity of 7 miles a second. The thermal energy of a gramme at this speed is 15,180 calories. . . . The energy of our most violent explosive—nitroglycerine—is less than 1,500 calories per gramme. Consequently, even had the explosive nothing to carry, it has only one-tenth of the energy necessary to escape the Earth—hence the proposition appears to be basically impossible."

Notice he doesn't dismiss the prospect out of hand but takes a moment to "critically examine" it. Then he follows the sort of routine sanity check that scientists do all the time and concludes, correctly, that the prospect is, currently, out of reach. In fact, it would remain so for more than thirty years, until 1959 when the Soviet Union launched the *Luna 2*, crash-landing it on the lunar surface near the Sea of Tranquility.

Bickerton *knew* that it couldn't be done; therefore he never asked himself, "*How* could it be done?"

Bickerton's knowledge of chemistry and ballistics was sound. His math was flawless, yet his conclusion was incorrect. He made an unnecessary, limiting assumption—namely, that any moon mission would be launched with a single, violent explosion, like a

32

bullet from a gun—the same method Jules Verne posits in his early science fiction novel, *From the Earth to the Moon*. But anyone who's ever watched a televised Apollo or Shuttle launch knows that's not the case. Liftoff is *slow*. Escape velocity isn't achieved until much later. Modern rockets achieve velocity by controlled combustion—continuous thrust rather than a single explosion.

Again, there's nothing inevitable about this trap. Many experts can and do work around it by remaining aware of their biases. Their attitude is what's important. The bias I call expert thinking is simply a trap to be acknowledged and avoided. I revisit the notion of making unwarranted limiting assumptions in chapter 3 on groupthink and on functional fixedness in chapter 4.

Finally, I want to take a quick look at how expert thinking can affect nonexperts. Faced with a hard problem outside their domain, nonexperts are too ready to surrender their own judgment to experts in the field. They assume, incorrectly, that if a solution were possible, the experts would know. But as we've already seen, experts are predisposed *not* to think this way.

I have a name for the epiphany someone experiences when they suddenly realize this: I call it *seeing the little dude*—the man behind the curtain. Just as Dorothy, after crash-landing in Oz, pinned all her hopes of getting home on the fabled wizard, only to learn in the end that there was no wizard, just a little dude hiding behind a curtain of authority, and that she'd had the ability to return all along, so we have to be careful not to cede too much authority to experts when pursuing disruptive innovation. You have to take responsibility for your own thinking instead of surrendering your thinking to the wizard—or to those around you who seem to know what they're doing.

If an elderly but distinguished scientist says that something is possible, he is almost certainly right; but if he says something is impossible, he is very probably wrong.
—Arthur C. Clarke

Experts are great at sharing what they know about what has been done. They're not as good at knowing what's possible that's never been done before. The trick is to benefit from their knowledge without being limited by their thinking. The wizard really is just a little dude behind a curtain and not the All-Knowing Oz he claims to be.

Most people don't care to see the little dude; they'd rather just follow along with what they're told. But to build a future that doesn't yet exist, you have to acknowledge that no one knows the answers. That's why there are no road maps at the tip of the spear. When you're heading into unknown territory, no one knows the way. When you're attempting to solve an unsolved problem, then, by definition, no one knows the solution. Deferring to experts not only limits your own chances at obtaining an answer, it almost guarantees that no one will find one.

For disruptive innovation, *there are no experts*. If the problem area were well enough understood for there to be experts, then the solution would not be disruptive. When you want a moon shot, you don't want experts who know it can't be done. Smart, educated, and motivated people working outside their safety zone lead to disruptive innovation.

Where No One Has Gone Before: Exploring the Unknown with Strong AI

> *. . . to explore strange new worlds, to seek*
> *out new life and new civilizations,*
> *to boldly go where no man has gone before.*
> —Captain James T. Kirk

Artificial intelligence (AI) is surging in popularity, but it's a limited sort of intelligence, "weak AI," that depends on access to very large amounts of data to accomplish what human beings can do with much less. Weak AI excels in specific, narrow problem domains but stumbles when it encounters anything truly new. To cope with the unknown, we need *strong* AI—what we call *anthro-noetic* (*noetos* is Greek for intelligence) AI, a general, flexible, more humanlike intelligence that's able to make decisions with limited data in ambiguous situations, as people do all the time. At Oceanit we're developing strong AI approaches to address a number of interesting but especially difficult problems, including malware detection, cancer, and in-flight medical assistance for a manned mission to Mars, among others.

Where No Man Has Gone Before?

One of the ironies of *Star Trek: The Original Series* is that, despite the crew's mission, when they landed on a planet they discovered that almost invariably, someone *had* been there before: Romulans, Klingons, Vulcans, or even some abandoned colony of old Earth. Even when they encountered new alien species, they were often distorted reflections of human beings, familiar but with a twist. That's because creator Gene Roddenberry was more interested in exploring contemporary social issues through a new lens than exposing a 1960s television audience to anything truly alien.

In late 2014, a group of politically motivated hackers exploited a *zero-day* vulnerability in the Sony Corporation's network security, obtaining access to sensitive internal documents, including executive emails and the digital prints of four unreleased films. The break-in is estimated to have cost Sony millions of dollars in damages and even more in negative publicity.

The hackers released 200 gigabytes of data and claim to have stolen more than 100 terabytes. For Sony, having their secrets revealed—confidential memos, salary histories, and more—rocked the company, leaving them exposed and vulnerable. If you've ever been the victim of a break-in, you know the feeling.

Zero-day attacks are nearly impossible to defend against because they exploit previously unknown weaknesses. The term "zero-day" means there are zero days between when an attacker exploits a vulnerability and the vulnerability is disclosed. The 2017 Equifax data breach, by contrast, was *not* a zero-day exploit. The hackers targeted a vulnerability that had been disclosed months earlier and had already been patched by the vendor in the latest version of the software.

Traditional antimalware software depends on identifying potential incoming threats against a database of known offenders. Zero-day exploits haven't been cataloged, so they can slip through.

The zero-day problem is one we're currently attacking at Oceanit with strong AI. Cybersecurity is becoming increasingly important as more and more sensitive data is stored online and as the sophistication and sheer resources attackers bring to bear increases. Some companies are now creating anti-malware software that relies on machine learning to analyze commonalities in existing malware in an attempt to identify new threats preemptively. And while they've had some success, we believe this approach will prove insufficient, as they'll never be able to identify genuinely innovative attacks.

Machine learning is an example of a big data-driven approach to artificial intelligence, what I call "weak AI" in contrast to the strong AI approach we're focusing on now at Oceanit. We've built weak AI solutions—neural nets, expert systems, machine learning, and more—for decades, but they have extremely narrow applicability and will fail on specific issues. Moreover, the limits of Moore's Law put an inherent cap on these approaches (see "Room at the Bottom" below).

An ideal weak AI system has infinite data, which is neither practical nor possible. That's why edge cases always cause problems for weak AI. A weak AI system for self-driving cars might recognize a person walking or riding a bicycle, but it will fail to recognize a person walking a bicycle unless that possibility is explicitly accounted for, thus putting the person walking the bike at risk.

So, to address the zero-day problem, rather than combing through massive data sets—since we can't just "look up" malware that has never existed before—we're developing a system that can assess a piece of software's intention as good or bad without ever having encountered it before.

People do this all the time with other people. For example, one Sunday morning I was running on the beach when I came upon a group of four big guys surrounding another guy waist-deep in the shore break. Depending on how you grew up, you might think a fight was about to break out. An algorithm trained to interpret men standing in these positions might determine that, based on proximity, number, and so on, that something bad was likely to happen. But if you also took into account that it was Sunday morning, that people on the shore watching didn't look like world wrestling fans, and that the onlooking children were dressed in Sunday beach aloha wear and not mixed martial arts T-shirts, you might conclude that this was a baptism and not a fight.

We evaluate strangers as potential threats, or not, without referring to a database. We don't consult a book of mugshots at every social encounter. Nor do we rely on a simple rules-based analysis. If you're with someone who intends you harm, you have a sense of that, even if you can't articulate exactly how or why. You could be wrong, but you make an assumption based on behavior, body language, and other cues. If someone points a gun at you while claiming to be your friend, you discount their verbal claim because of the threat presented by the gun. You read their intention, not the words they say.

But couldn't a computer be trained to look for a gun? Of course it can. That's one scenario handled. But what about a knife? A chain? What about a pipe wrench or a baseball bat? How do you get a machine to distinguish between a thug who wants to break your legs and a coach who wants you to join their softball team? Looking for a weapon is rules-based thinking: machine thinking. People are more flexible. They excel at making decisions in ambiguous contexts with limited information. Machines can't—yet.

Because of our strong AI approach, we were selected by NASA as part of their iTech initiative to explore the possibilities of virtual medical staff to monitor and assist with crew health on a manned mission to Mars. We were up against a number of competitors, all with strings of successes, but all dependent on a data-driven approach to the problem. We successfully argued that for the first manned journey to Mars, there *is* no data to extrapolate from since humans haven't been to Mars yet. You'd want a smart, small-data system that can make intelligent decisions without depending on prior knowledge.

Another area of application is virtual office staff, using strong AI to read emails and summarize them, pulling out the important information. It's a hard problem. To do it well, you need to understand language. We're evolving a system to do that. Right now, we have a system that understands things at the level of an infant, and we need to progress that to the level of a college student to be useful. Again, this is a general approach, which, when it works, will be applicable to any domain. Google is currently attacking the problem from the other end, extracting specific information in narrow domains from your email, like flight and hotel booking information. The Google approach is more immediately useful—it's in place now—but it will be difficult to scale effectively.

Room at the Bottom

We got started on strong artificial intelligence, what we refer to as anthronoetic or humanlike AI, by asking ourselves about "headroom" on processor development. For a long time, Moore's Law, the famous claim that microprocessors would double in speed every eighteen months, swallowed up a lot of software sins. Your program's not fast enough? Don't bother to rewrite it more efficiently, it'll be plenty fast on the next generation of hardware.

But there's a physical limit on how small a transistor can be, and we're rapidly approaching the end of the road for

Moore's Law. Today Intel produces a chip in a 9 nm process (where the size of a physical feature on the silicon is nine nanometers), and IBM has demonstrated a transistor on the 3 nm scale. But the size of a silicon atom is just 0.2 nm, putting a hard cap on how many more doublings are possible. At some point soon, software won't be able to coast on the free ride of faster hardware. Software will have to get smarter.

After the initial promise of artificial intelligence in the 1980s, including Japan's fifth-generation project, the field lost momentum, experiencing what's sometimes called "AI winter." AI researchers complained about shifting goalposts in the definition of intelligence. When a previously unthinkable milestone was achieved—a computer program beating a human grandmaster at chess—detractors said, "But that's not *real* intelligence. You're just brute-forcing the solution space." The term "artificial intelligence" fell out of fashion, as researchers abandoned the goal of mimicking human intelligence and instead referred to specific technologies associated with the field: expert systems, neural nets, machine learning, and so on. All of these, by the way, I lump together as weak AI, or data-driven AI, if you prefer a less pejorative term.

I'm not denigrating weak AI. We've been using it ourselves for over thirty years. Sometimes, particularly in a limited, well-understood domain, it's the right approach. IBM's Deep Blue defeated Garry Kasparov, and its Watson program trounced two Jeopardy champions in a contest of general knowledge. At Oceanit we just completed a project for an oil company where we created an autonomous drilling solution that performed within a percentage point of an expert human driller. In that case we simply created a high-quality neural net that worked well enough to get the job done. But none of these programs can generalize. Deep Blue doesn't play poker. Our drilling program can't pilot a ship. And Watson, despite its sophisticated natural language interface, knows only what it has been taught. For all its successes, Weak AI is inherently limited by what it already knows. If we want to tackle broader, fuzzier domains, we need a different approach.

I believe strong AI will succeed where weak AI can't by doing more with *less* data. Rather than looking up the answers, strong AI deduces them from fundamental principles. Rather than looking for correlation, it looks for causality and explanations. Rather than looking for patterns, it tries to make predictions. This gives it the flexibility to deal with novel situations, just like a human being.

So how do we divine the intention of software? We could defer judgment until we can evaluate its behavior. But then, if it proves malicious, it would already be too late. We'd like to operate more like *Minority Report* and intercept the bad behavior before it even happens. Software's behavior is dictated by its code, and we have access to that before we ever have to run it, allowing us to predict the future. Solving zero-day creates a "pre-crime" capability.

Because we believe that human intelligence is built around language, we've been studying the mathematics of language and analyzing the software from that perspective. Essentially, we're bringing natural language processing tools to bear on suspect code and "reading" it to see what it's planning, what its intentions are.

The fascinating thing about this approach is that we think we can apply it in medical contexts as well, to read the intentions of genomes and cells. We have a lot of cruft in our DNA—leftover stuff that was important at some stage in our evolution but is unnecessary now. Sometimes these genes get turned on and create cancer or other problems. Currently we don't recognize disease until it manifests. We'd like to be able to intercept bad cells or bad genes in advance, to prevent the disease before it even happens, the same way we're planning to intercept malware.

> *An expert is one who knows more*
> *and more about less and less*
> *until he knows absolutely everything about nothing.*
> —Nicholas Murray Butler

Remember those analogy tests from your SAT?

> Kitten : Cat ::
> Puppy : Dog

That reads, in case you've forgotten the format, "A kitten is to a cat as a puppy is to a dog." Weak AI is terrible at these, by the way. Well, here's an analogy for you:

> Intellectual Anarchy : Strong AI ::
> Traditional development : Weak AI

Or in other words, Intellectual Anarchy is to strong AI as traditional development is to weak AI. Strong AI isn't strictly superior to weak AI, just a different approach to problem solving that promises to be much more effective in uncharted territory. So, too, Intellectual Anarchy isn't strictly better than traditional development, unless you need something revolutionary, when you're going where no one has gone before. Each has its area of application. Oceanit has a long tradition of employing weak AI methods in autonomous systems (like the aforementioned drilling system), sensor integration, and short-term weather prediction via 3-D ceilometry, but strong AI is better suited for the types of hard problems we like to tackle. Similarly, we rely on aspects of traditional development methodology in addressing interesting and challenging problems, but Intellectual Anarchy allows us to go after truly disruptive innovation.

This approach is reflected in the type of people we like to work with. We employ a lot of PhDs. Unlike other companies, we typically don't hire them for their domain knowledge. We're not looking for experts, but for people with the proven ability to solve hard problems. But that's just the beginning. We also need them to be agile and flexible, able to work in multiple areas, able to work with limited data and be comfortable with ambiguity—the same qualities that differentiate strong AI from weak AI.

Experts are great when you can point them at a problem they already know how to solve, but many of the problems we tackle at Oceanit have no obvious solution—they're unsolved. Until we get started on them, we don't know how, or if, we'll be able to solve them. We need people who are comfortable operating outside their area of expertise. If you think of generalists as being functional in many areas but without great depth of knowledge in any one area—wide but shallow—and specialists as the opposite—narrow but deep, then our best people at Oceanit tend to be "T-shaped," combining broad general knowledge with depth in a particular field.

I take a deeper look at what makes a great fit for Oceanit in chapter 8 on Techno Warriors.

///////////

Implications

There are basically two failure modes in expert thinking that diminish disruptive innovation, that broad category of problem-solving where most would agree there is really no obvious path forward. The first failure mode comes from surrendering one's ability to critically question the expert, the "man behind the curtain." The second is not recognizing when that expert is at risk of motivated reasoning, earned dogmatism—the right to remain close minded. Methods and materials are constantly changing, and moving faster each year, so what was once impossible may become possible. Nevertheless, the expert may not see it that way. The risk is becoming part of this expert-thinking cluster of potential cognitive biases that can affect just about anybody, even scientists and researchers. But to address the most important challenges, one needs to be willing to navigate the boundary between the known world and the world yet to be discovered.

If there's one area where experts are stumped and bunkering down in dysrationalia (defined as the inability to think and behave rationally despite adequate intelligence), it's the future of artificial intelligence. The implications of AI are on a scale so vast that they're difficult to comprehend. Thinkers from I. J. Good to Verner Vinge to Ray Kurzweil have put this incomprehensibility at the center of their definition of the technological singularity, the moment when our technology coalesces and compounds to a point where we create superintelligence. What happens after? No one can guess—it's a singularity! It's like the event horizon of a black hole, beyond which no light or information can escape.

A century ago, industrialization and automation were perceived as threats to workers, yet the world of work evolved with new industries and new needs. Now we're confronting similar issues with artificial intelligence. Just as humans, society, and business both influenced and were changed by the Industrial Revolution to the eventual benefit of all, I believe AI will be a net positive.

I'm optimistic about what AI can bring to the world because science and technology have granted so many incredible gifts of health, wealth, knowledge, and freedom to humans and society. Just one startling example can stand for the myriad benefits of scientific knowledge: science eradicated smallpox, a painful and disfiguring disease that killed 300 million people in the twentieth century alone. This awe-inspiring achievement alone should put the lie to any moaning that we live in an age of decline, disenchantment, or meaninglessness—as should the lives of the 300 million-plus people who won't die of smallpox in this century. General AI is expected to have an impact on humanity of this scale or greater.

We can group AI implications into three buckets as follows:

1. *Human Assisted Tools:* This is the form of simple, limited AI we've been living with, quite successfully, so far. It has improved productivity, reduced monotonous, repetitive labor, reduced costs, and improved quality in simple mass production tools and in the application of clever algorithms that simplify calculations and analysis. In the auto industry, it has improved everything from antilock brakes to automobile performance and safety diagnostics.

Advanced algorithms control systems for sorting fruit, counting packages, and trading stocks. When we ask Siri to find a particular restaurant, or Google to find an affordable car, as long as the task is very similar to the AI training environment, it all works. This type of AI is extremely limited; it's easy to make it work only for very narrowly defined, specific applications. But this type of AI fails when challenged with edge cases—first-time encounters with limited experience, something humans easily handle. We have seen this in the tragic results of AI in autonomous vehicle testing.

2. General Artificial Intelligence / General Universal Intelligence: This type of "strong" AI is on the horizon. It is both scary and inspirational. Many hyperbolic statements have been made regarding artificial intelligence, as it goes through the Gartner hype cycle, from emergence and early adopters through to scaled use across many industries. While IBM's Watson can beat a human at Jeopardy and Google's Alpha Zero can beat a human at Go, we are still a long way from a truly, humanlike AI. However, "strong AI" has the potential to create new understandings of everything from cancer to space travel.

General intelligence will be able to help solve problems that seem unsolvable in health care, disease, and longevity, as well as in developing new material and methods in all the sciences. Our work at Oceanit in strong AI, on understanding causality in biology, for example, could lead to a different approach to eradicating cancer.

Science fiction introduced us to an out-of-control HAL in *2001: A Space Odyssey*. A true general AI will have to include morality and other human values. If strong AI is a new type of expert, it will be interesting to see if it, too, is susceptible to ego-driven motivated reasoning and biased decision making, or if, free of such human traits as pride or embarrassment, it will be more forthcoming than we are about its own limitations and abilities.

3. The Future of Work—and Everything Else: This is probably the biggest change humans will face in the next century. When we can solve just about anything, we will be confronted with some philosophical questions: What should we do with our lives and our time in the world? What's the purpose of education? How does society function when work no longer requires a hierarchy of skills based on education? What, exactly, will the word "economy" even describe in this possible future?

I believe the future of work will yield jobs where "work and play are indistinguishable"—jobs that may not be imaginable today. At Oceanit, we want some aspect of work and play to be indistinguishable—that's part of Oceanit's mission statement, developed back in 1985. It's clear to me that this merging of "work and play" makes life exciting and rewarding. Although this eventuality will take some time to be available to everyone, it's a possible end state for humans when general AI becomes mainstream.

In the mid-twentieth century, Abraham Maslow introduced and perfected his theory that humans have a hierarchy of needs, from physiological (food, clothing, shelter, sleep, and more) to safety (health, finances, emotional security, physical safety) to social belonging, then self-esteem, self-actualization, and finally, transcendence. As Maslow defined it, "Transcendence refers to the very highest and most inclusive or holistic levels of human consciousness, behaving and relating, as ends rather than means, to oneself, to significant others, to human beings in general, to other species, to nature, and to the cosmos."

///////////

Our economy-shaking revolutions seem to be moving up through the same hierarchy. The Agricultural and Industrial Revolutions have brought down the costs and vastly expanded the accessibility of such basic physical requirements as food and shelter. The Information Age, the Internet, social media, and more engage needs as far up the hierarchy as self-esteem and self-actualization; not perfectly, by any means: the bullying that happens on social media comes hand in hand with the self-expression and the ability to help people organize spontaneously for good, such as Gofundme campaigns for people in need. What would an economy, a society, look like if humanity need only concern itself with transcendence?

CHAPTER 3

Defeating Groupthink:
Diversity of People, Projects, and Place

Whenever you find yourself on the side of the majority,
it is time to pause and reflect.
—Mark Twain

My first job out of college, and my first major encounter with *groupthink*, was with the Storage Technology Corporation (STC) in Colorado. Founded by former IBM employees, STC initially produced tape drives and media to service those customers IBM abandoned when IBM discontinued tape storage in favor of disk storage. Later STC moved into disk storage themselves as an upgrade to their customers.

STC hired me right out of college as an entry-level process engineer. I'd just graduated with a degree in engineering physics from the University of Colorado in Boulder, paying my way by working in university labs in the winter and running my own landscaping and irrigation business in the summer. When STC hired me, it was the realization of a dream. I was the first person in my family to graduate from college and I'd just landed a prestigious technology job. It actually paid less than what I'd been making putting myself through school, but I knew technology was the future.

STC put me to work on thin-film magnetic heads for their new hard drives—the read-write heads that "flew" mere microns over the massive yet then state-of-the-art, 18-inch aluminum-oxide disk platters. The technology was so new that I was assigned to their research and development (R&D) group, working out of the Electrical Engineering Department at CU Boulder.

As part of R&D, I felt the need to share my ideas, particularly the idea of building a personal computer that could sit on a desktop, rather than a mainframe. That way one could do all types of calculations and analysis at one's leisure, versus waiting for time on a large system that is typically reserved for special use. That was a big deal back then, since IBM punch cards and batch computing were the norm. But I knew it was possible, because in high school I'd built a simple binary calculator with vacuum tubes.

My bosses, those same former IBM employees, told me it wasn't possible. They told me, "Young man, even if you could, there's no market."

Now if they'd told me that it didn't fit with their strategy, that the small computer market was too volatile, that they feared the entry of a major player like IBM, then so be it. STC wasn't wrong to resist building a personal computer. The company still exists today (as StorageTek, now anchoring the tech corridor between Boulder and Denver), unlike a lot of companies that dove headfirst into the personal computer revolution.

At the time I attributed their skepticism to the fact that the team leader had a PhD and I didn't. I wasn't credible. That was part of my inspiration for returning to school and getting my doctorate. But looking back, that was my first major encounter with groupthink. STC was so entrenched in the mainframe industry that they couldn't see the revolution about to erupt under them. I wasn't familiar with the term "groupthink" at the time—all I knew was that I didn't belong there; I was going to grad school.

Groupthink is a cognitive bias that impairs decision making by groups of otherwise highly qualified individuals. That is, a group of smart people in a room will sometimes make a worse decision than any of them would make on their own. The term, a deliberate evocation of Orwellian "newspeak," was coined by social psychologist Irving Janis in 1971 as a label for the phenomenon whereby social pressure can lead to a deterioration of "mental efficiency, reality testing, and moral judgment" in small groups of government policy makers. The term has since evolved to include other organizations, particularly businesses, and groups as large as national political parties.

STC wasn't alone in their casual dismissal of the potential of personal computing. Many mainframe and minicomputer manufacturers were slow to recognize the potential of personal computing, and we can recognize many of Janis's groupthink symptoms in their dismissal: the perception of invulnerability, feelings of moral superiority, the stereotyping of outgroups (personal computing was regarded as something for hobbyists, not serious business), and so on.

A more tragic example of groupthink was the decision to launch the space shuttle *Challenger* on January 28, 1986, at unprecedentedly low (near freezing) temperatures. The O-rings (gaskets) that prevented hot gas from escaping the solid rocket boosters, and which weren't rated to be effective below 53 degrees, failed. The shuttle exploded seventy-three seconds into its flight, killing all aboard, including Christa McAuliffe, an elementary school teacher selected as a "Teacher in Space."

In the wake of the disaster, investigators learned that engineers voiced multiple safety concerns, but they were overruled by management or not passed up the chain of decision making.

Go Fever: Groupthink at NASA

NASA has its own term for the particular flavor of cognitive bias that surrounds a launch. They call it "go fever." But the space shuttle *Challenger* disaster fits almost perfectly into the framework that social psychologist Irving Janis lays out for groupthink: invulnerability, rationalization, moral superiority, stereotyping, peer pressure, self-censorship, assumed unanimity and mindguards. Almost all of these symptoms were present in the flawed decision making leading up the launch.

On the day of the *Challenger* launch, there were multiple safety concerns: the shuttle had never launched in such low temperatures, there was ice on the launchpad, and the O-rings had been exhibiting unexpected erosion. All of these concerns were dismissed.

The presidential Rogers Commission ultimately found that the failure of the O-rings was responsible for the destruction of the *Challenger*. Quotes below are from the *Rogers Commission Report*, which adopted the Janis list of primary symptoms of groupthink:

- *Invulnerability:* The O-rings were observed to take damage during flights, noted as a possible flight risk, but this concern was waived each time. "NASA and Thiokol accepted escalating risk apparently because 'they got away with it last time.'"
  Commissioner Richard Feynman said, "[The Shuttle] flies [with O-ring erosion] and nothing happens. Then it is suggested, therefore, that the risk is no longer so high for the next flights. We can lower our standards a little bit because we got away with it last time."
- *Rationalization:* The O-rings work in pairs, and the assumption was that even if the first O-ring of the pair failed, the second O-ring would prevent a blowout. This was an untested assumption, and it was directly against NASA's safety protocol. The O-rings were a so-called Criticality-1 component: they are not allowed to rely on redundancy.
- *Peer pressure:* At a preflight conference to determine if the low launch temperate was a concern, every engineer present from Morton Thiokol (the contractor responsible for the O-rings) argued that it was. Management initially agreed, then reversed their position. The commission concluded that "Thiokol Management reversed its position and recommended the launch of 51-L, at the urging of Marshall and contrary to the views of its engineers in order to accommodate a major customer."

- *Self-censorship:* Multiple people involved in the launch, though they voiced their concern, allowed themselves to be overruled.
- *Mindguards:* Information critical to the safety of the launch was sequestered at lower levels, rather than being passed up the chain to the ultimate decision makers. From the report: "The Commission is troubled by what appears to be a propensity of management at Marshall to contain potentially serious problems and attempt to resolve them internally rather than communicate them forward."
- *Assumptions of unanimity:* Because top-level decision makers were not informed of several serious risks, they assumed conditions were safe when in fact they weren't. From the report: "If the decision makers had known all the facts, it is highly unlikely that they would have decided to launch 51-L on January 28, 1986."
- Ironically, the Rogers Commission, assigned to investigate the disaster, nearly succumbed to groupthink itself. Their initial tactic was to question the managers, not the engineers—the same managers who had overruled the engineers and suppressed critical safety concerns. Only the presence on the commission of Richard Feynman, who insisted on talking to the engineers directly, led to the discovery of the O-ring problem (a "discovery" that was well known at the time of launch).

///////////

Groupthink exists for a reason. Humans are social animals. Our instincts guide us to trust the group, often for good reason. Or, as Randall Munroe more cleverly puts it in his XKCD cartoon, in response to the perennial question about following your friends off a bridge, "Which scenario is more likely: every single person I know, many of them levelheaded and afraid of heights, abruptly went crazy at the same time . . . or the bridge is on fire?" Yet those same instincts can also lead us astray.

An ironic consequence is that the groups most at risk for group-think are those that otherwise seem to be functioning at a high level—with high group cohesion, friendships within the group, and little conflict. With dedication to a shared vision and little friction, groups can make rapid progress toward their goals. Where they fail is evaluating whether those are the *right* goals. What Janis calls "the nondeliberate suppression of critical thoughts as a result of internalization of the group's norms" can be an advantage in many situations. It's the "nondeliberate" part we have to be careful of.

Where groupthink absolutely fails is in fostering innovation. In a changing world, every business must innovate, even those that don't have innovation as their focus. They must innovate to survive. Every business evolves to meet the conditions in which they operate—the availability of certain people or technology, the current market demand, or political pressures. Policies and procedures are put in place to streamline operations for that environment, but those same policies and procedures can become calcified—locked in. People forget that they were created for specific conditions and assume that is just the way things are done. Groupthink creates an inflexibility and resistance to change. When change inevitably occurs and those same businesses fail to adapt, they suffer.

For example, Kodak, which invented digital photography in 1975, was so protective of its traditional film business that they never capitalized on their enormous technical lead, eventually filing for bankruptcy in 2012.

For businesses that *do* center around innovation, those responsible for generating new ideas and products, the situation is exacerbated. Innovation depends on a diversity of ideas and opinions. Groupthink reinforces the opposite, creating a culture of orthodoxy. New ideas, especially "impossible" ideas, are easily shot down by conformists and mindguards, or worse, never voiced at all due to self-censorship.

At Oceanit, disruptive innovation is our raison d'être. I wasn't familiar with groupthink when I started the company on a shoe-string in 1985, but as the company evolved through trial and error—keeping what worked, discarding what didn't—I always tried to optimize for innovation. The result is a company that is structured to avoid cognitive biases like groupthink and expert thinking.

There are four main strategies that help defeat groupthink at Oceanit. Three of these came for "free," in the sense that they fell out of an imposed condition (geography) and best practices with a completely different focus (hiring for excellence and creating space for feedback). The fourth came later and was specifically aimed at reducing one aspect of groupthink (an in-house, inspirational lecture series called No Limits).

First, our location in the middle of the Pacific, far from any of the usual tech hubs, removes us from the industry-wide groupthink that infects such places. First-time visitors and clients are sometimes surprised we can be as effective as we are when we're 3,000–6,000 miles removed from anywhere "important," but we see it as an advantage. We're not driven by the need to comply with VC-think and the tech media, to pursue hockey stick growth, or to fall into the billion-dollar, valuation-focused, unicorn-or-bust mentality so prevalent in Silicon Valley, for example. (There are hosts of other advantages that result from being outside the mainstream. I have much more to say about this in chapter 5: "Defeating Geography.")

Second, we create space for feedback. This is not occasional sporadic feedback from suggestion boxes and hallway conversations—opportunities for employees to suggest ideas, voice concerns, and share their perspectives are baked into our operational calendar. In the fourth quarter of every year, we hold our annual strategic review. Anything and everything is open to discussion, from individual projects to administrative procedures to entire lines of business. We open everything up and then we lock it down and operate on it for a year. A business can't run on *constant* change. Here are some examples that came directly from this feedback process, from minor to major:

- *Paperless receipts:* Not the most radical innovation, but an example of an operational change that came at the request of our road warriors at one of our fourth-quarter meetings. They were frustrated at the accounting department's insistence on paper receipts for travel expenses (to satisfy IRS requirements). We found an electronic solution that satisfied both groups, streamlining the process for everyone. When you have more people untethered to the office, you have to find a way for them to be more efficient.

- *Human-centered design:* The technology we develop needs to impact humans and society, so we resolved to develop our expertise in human-centered design, not because it was a natural strength but because we felt it was important. We often choose to take on challenging problems without necessarily having a clear way forward because the benefits outweigh the risks (and the way forward often reveals itself once we get started). We don't always succeed, but we do learn a lot, succeed or fail. In this case, human-centered design has become a core pillar of our methodology.

- *"Design Thinking" area:* Design Thinking has been a huge win for us, so we wanted to create an environment to support it, particularly since we believe it's essential to transition disruptive innovation from the lab to the market. We created a special area of the office where we got rid of all the desks and created mobile workstations—sofas on wheels—to create a more fluid work environment suitable for quickly assembling teams to design and build things. Everyone involved could quickly and easily reconfigure the layout to suit the task at hand, whether it was to pull one table away from the others for a solo project or quickly assemble an ad hoc group for a rapid proof-of-concept. (See chapter 11 for more on Design Thinking.)

- *Energy:* Although we've been involved in renewable energy for decades (including ocean thermal energy conversion, photovoltaic, and hydrogen), we made a decision to enter the oil and gas industry because it seemed important, not because we had any particular expertise. The original impetus was our concern over fracking and the contamination of the groundwater. Although we didn't have a solution in mind, we felt we could make an impact by attacking the problem. This led to our work with Shell, which is ongoing, and other major, global companies (see Anhdyra later in this chapter and the section titled "The First Thousand Feet: Secure Drilling with SCIN" in chapter 12 for an extended discussion).

Third, our hiring practices have led to a diverse, creative, and inspired workforce. People like to hire other people like themselves—it's natural; diversity can feel threatening—but that reinforces groupthink. We deliberately look for high-talent individuals we can put into the mix to stir it up, people who aren't afraid to have an opinion that may make others uncomfortable. You want that. You want to respect other people as individuals with different perspectives and experiences, but you don't want people too comfortable with their ideas of what works and what doesn't, what's possible and what's not. You want to foster debate and discussion.

We do tend to favor recent PhDs (for reasons I discuss elsewhere), which are a pretty selective group, but within that group we want as much diversity as possible. Demographic diversity, sure—we want people that don't have kids and people that have a big family, people who went to private school or public school or were homeschooled, people of different ethnic backgrounds, gay and straight, who come from different socioeconomic backgrounds—but more than these, we want people who think differently, with unique life experiences and outlooks. We look at their *play history*—what did they do when they were kids? What do they like to do in their free time? Different experiences bring different perspectives.

We hired a Stanford graduate, originally from India, an excellent candidate by any measure, but the thing that pushed him over the top for us was that before starting graduate school he took some time and built his own bicycle and rode it across Namibia in southwest Africa. That spoke not only of a level of independence, resourcefulness, and can-do attitude that are vital to tackling hard problems but was also a unique experience that provided a different perspective on the world. He has since helped us to connect human-centered design to deep science, enabling us to push back the edges of scientific knowledge but also to understand what it takes to bring that knowledge to society.

Another guy we hired got his PhD here in Hawaii. He was a mechanical/materials/nanotechnology engineer but also an incredibly prolific thinker and collaborator. He changed the conversation at Oceanit on how materials could be coaxed to behave. At nanoscale, materials behave in ways that defy our Newtonian physical intuition. The nanoworld runs by different rules. (See the sidebar on nanotechnology in chapter 7). Just getting people here to understand that rules change was a big step, but for those of us who live on the jagged edge of science, it was an exhilarating contribution.

A young guy originally from Hawaii we brought in to work on a machine-learning project had a background in linguistics and math at MIT but not machine learning and computation. But he broadened the conversation around artificial intelligence at the company. We'd been running into limitations with our machine learning, neural-net approach, and we shifted to our new approach, what we call anthronoetic AI—artificial intelligence designed to think like a human, not like a computer. This typifies our transdisciplinary thinking, but the key is that diversity isn't an end in itself but a means to an end. What we care about is the unique perspective that leads to new ideas. You have to seek that out wherever you can find it. You can't be afraid of people who don't think like you do, who don't have your background or shared experience.

Fourth, we encourage risk taking and individual action, especially in our newer employees, who can feel overwhelmed or intimidated at first or simply play it too safe.

A few years ago, we started our No Limits speaker series, bringing in individuals who inspired us with their boldness and vision to serve as models and inspiration. We learn from them the human story of disruptive innovation: the eclectic nature of discovery and the passion for life that we find is a common thread for those who drive to the edge of knowledge and disruptive thinking. Speakers have included the following:

- *Earl Bakken*, founder of Medtronic, on the future of medicine in Hawaii. Bakken, a true transdisciplinary thinker, pioneered modern electrophysiology with the invention of the pacemaker. Bakken was willing to follow his interests and instincts into areas before they were considered serious subjects.
- *Dean Kamen*, American entrepreneur and inventor, on fostering innovation. Kamen is a prolific inventor whose credits include the first insulin pump and portable dialysis machine, but he is most famous as the creator of the Segway. Kamen is a fearless innovator and demonstrates that you can innovate anything.
- *Roz Savage*, environmentalist and adventurer, on tackling overwhelming challenges in unpredictable conditions. Savage is the first woman to have rowed solo across three oceans (Atlantic, Pacific, and Indian), spending 500 days at sea in a 23-foot boat. Savage abandoned a comfortable life as a consultant to do something outrageous by asking herself, Do you want to be safe, or do you want to make a difference?

It took a conversation with the assistant secretary from the Department of Defense to make me fully appreciate how fortunate, and unusual, we've been to sidestep the groupthink trap. On a visit to Oceanit, he cited, as one of our chief points of interest, that we were "outside the Beltway groupthink." He was investigating the possibility of supporting more research outside of Washington for just that reason and saw what we were accomplishing in Hawaii.

"You have a whole different way to look at the world," he said. "I've never seen anything like it."

Ultimately, we defeat groupthink by remaining relentlessly focused on innovation. Everything stems from that: from our choice of problems to the people we hire to tackle those problems to the environment we create to support them in their tasks. Much of what we do at Oceanit we stumble on through a disciplined process of trial and error. Because we are so focused on innovation, and groupthink is so inimical to innovation, our evolution naturally took us away from that particular failing.

Anhydra: Hydrophobic Surface Treatment

If you do what you always do,
you'll get the results you've always gotten.
—Anonymous

The basic idea of nanotechnology was advanced by Caltech physicist and Nobel laureate Richard Feynman in 1959, who suggested that individual atoms and molecules could be manipulated and engineered. Swiss Physicist Heinrich Roher was awarded a Nobel Prize in 1986 for developing a scanning tunneling microscope, and that opened the world to nanotechnology. In 2006 Oceanit, together with the American Society of Mechanical Engineers (ASME) and the University of Hawaii, cohosted the Multifunctional Nanocomposites International Conference and NanoVenture Competition. This brought the world's leading experts in nanotechnology to Hawaii, including the 1996 Nobel laureate in Chemistry, Sir Harold Kroto, who shared the Nobel Prize with Richard Smalley and Robert Curl.

Nanotechnology
The term "nanotechnology" was first popularized by K. Eric Drexler in his book, *Engines of Creation: The Coming Era*

of Nanotechnology, in 1986. Drexler envisioned nanoscale robots called assemblers that could create literally anything imaginable, constructing things atom by atom. Corresponding disassemblers could reduce any substance to its component atoms. He also warned of a cataclysmic scenario where runaway, self-replicating assemblers/disassemblers reduced the entire world to "gray goo."

This vision of nanotechnology took the science fiction world by storm, fueling near magical tech in novels, TV, and film. In Neal Stephenson's *The Diamond Age,* every home has the Feed, supplying raw materials from which anything can be created, as needed, by Matter Converters, then recycled when finished with. Greg Bear's *Blood Music* features a bio-tech spin where nanotechnology is used for human augmentation but ends badly. And in Michael Crichton's *Prey,* swarms of nanobots become self-aware and turn deadly. The Borg from *Star Trek* and the T-1000 from the *Terminator* series are also animated by nanotechnology.

As thrilling/terrifying as this conception of nanotechnology may be, it's a long way from the current state of the art, and the term has been generalized to refer to any engineering that takes place on a *nano* scale (a nanometer is 1×10^{-9} meters, or one-billionth of a meter). Experimentalists are discovering new nanofeatures and qualities every day.

Nanotechnology is engineering in the small—the very, very small. More than just another buzzword, nanotech is engineering on a game-changing scale where the normal rules no longer apply. *Nano-* is the metric prefix for one-billionth in the International System of Units, just as *micro-* is the prefix for one-millionth and *milli-* is the prefix for one-thousandth. Microscale engineering is already commonplace. Apple's iPhone, for example, is manufactured to tolerances as tight as one micron (one-millionth of a meter). To put that in perspective, a human hair is about 15 microns wide, and an *E coli* bacterium is about 0.5 microns across. A nanometer, on the other hand, is one thousand times smaller than *that*.

Our normal, physical intuition fails us at this scale. Newtonian physics breaks down and quantum effects take over. Materials behave in counterintuitive ways and become very hard to reason about. The work becomes less theoretical and more experimental. You have to try things and see what happens. That's what makes it interesting.

After the conference, we invited Sir Harold to Oceanit as a No Limits speaker. We asked Sir Harold to share his experiences and inspiration that led to his Nobel Prize, which included his passion for soccer and art. Kroto is best known for his role in discovering *buckminsterfullerene* ("buckyballs"), a 60-atom carbon molecule in the shape of a soccer ball, with potential applications in communications, medicine, and energy generation and storage. He was inspired by art, sports, and other things that led him to find a connection between chemistry and the rest of life. He showed that having outside interests and hobbies enriched his ability as a serious scientist. You can see his presentation at https://youtu.be/ssydy-lvw7Q.

Although at the time we had limited experience with nanotechnology, we agreed that it was interesting and important, as well as worth taking the time to explore and try to understand how it could impact humans and society. Cohosting this conference had introduced the Oceanit team to other world leaders in nanotechnology, and we got to know several personally, some of whom we have collaborated with ever since.

To help the Oceanit team get comfortable with the nano world, we decided to build a nano surfboard by mixing titanium nanotubes with surfboard resin, which made the surfboard much stronger and more resistant to dings but also maintained its light weight and high performance, as discussed in another context in chapter 7. That exercise opened our eyes to a big gap in the nanotechnology world: industrial nanotechnology, which requires durability as well as performance in high temperature and harsh environments. So, to pursue this subcategory of nanotechnology, we persuaded the US Navy to fund a multiyear effort to consider how industrial nanotechnology could improve shipyard performance—upgrading materials and methods to enable a more competitive, more responsive, and lower-cost shipyard.

In the process, we experimented with many, many ideas. One of them became Anhydra, a semidurable nanosurface that repelled water, a property referred to as "hydrophobic." This was an engineered surface inspired by the biomimicry of a lotus leaf, which has a self-cleaning, hydrophobic surface. Looking through a microscope, you find that the surface of a lotus leaf is not really smooth, it actually has many pointy surfaces, like a bed of nails. Just as with a bed of nails, if you lie on an individual nail, it might go through you, but if you lie on the whole bed of the nails you can stay on the surface without being impaled. Hydrophobic nanosurfaces work similarly: water beads up on the pointy nanosurfaces, forming an intriguing and beautiful ball of water. With this biomimetic inspiration, we engineered the surface of a piece of metal by ripping metal on a nanoscale to create a metallic lotus leaf–like surface. This worked remarkably well to repel water. After this remarkable finding, we then asked ourselves: So what? Who cares, and what does this mean?

Without a clear application, Oceanit's Tech Sherpa developed a silly game of water hockey, where the ball of water on the surface acted like a hockey puck. Two players can hold onto opposite ends of a flat piece of metal to see who can move the water ball puck into the goal. We weren't sure what to think about the superhydrophobic surface, but we thought perhaps there might be a consumer game or some other application we hadn't yet imagined.

Oceanit has a very active summer intern program and has hosted close to 600 interns over many years. We receive close to 1,000 applications for ten to fifteen slots per year. One of our summer interns was going back to college where she was a senior in mechanical engineering. Like most ABET–accredited engineering programs, she was required to propose engineering research as her Senior Project. This intern asked if she could use the Anhydra material to measure precisely how much drag is reduced by the Anhydra surface treatment. More specifically, she asked this question: How much would it reduce the drag and increase the flow of water if Anhydra were on the inside surface of a pipe or tube? The result caught us all by surprise: it turns out that it reduced the drag significantly, up to 35 percent. With this interesting finding, we decided to see who in an industry that relies on pipes might be interested in this insight.

As noted earlier, we had made the decision at Oceanit, prompted by one of our annual strategic reviews, to get involved in the oil and gas industry—not because we saw a particular opening but simply because we felt it was interesting and important. In this case, our immediate concern was the threat of groundwater contamination presented by the rise of fracking, but we felt there must be multiple opportunities for disruptive contributions in such a massive industry. While we didn't have any relevant expertise, we felt our outsider's perspective could prove valuable.

The oil and gas industry has been dominated by groupthink for decades. The major players (Exxon, Shell, Chevron, British Petroleum, etc.) all rely on the same technology, delivered by the same small group of service providers, with the consequence that innovation in the industry has stagnated. Haliburton supplies everyone with the same cement. Schlumberger provides the same downhole technology. Baker Hughes offers the same tools. Change is very slow, absent a new bell or whistle. Why change what's not broken? Thus, the service providers, not the oil companies, have been setting the boundaries of what's possible in the industry. As the oil companies have exhausted their reserves of easily accessible oil and pushed into more extreme environments (deep wells, offshore drilling, and fracking), their technology has failed to keep pace with the demands for affordable, safe, and environmentally sustainable extraction. These companies are not incompetent. In fact, they're extremely capable, and the engineering challenges they tackle are enormous, but they've fallen into a fixed mindset.

Recently they were dealt a pair of wake-up calls in the form of the Macondo blowout in the Gulf of Mexico (subject of the recent film *Deepwater Horizon*) and the commercial development of fracking.

The Macondo blowout—an explosion on the Deepwater Horizon rig, drilling in 5,000 feet of water at the Macondo Prospect—killed eleven crew, injured seventeen more, and resulted in one of the greatest environmental disasters in history, spilling the equivalent of 5 million barrels of oil into the Gulf. The investigation into the cause of the accident reads like a replay of the space shuttle *Challenger* explosion, with many of the same hallmarks of groupthink as identified by Janis—including taking insufficient precautions in extreme conditions, violating best practices (by drilling in the presence of methane deposits), and ignoring the safety concerns of engineers on-site.

The Macondo well stretched nearly 18,000 feet below the seafloor, far from a record for the Deepwater Horizon rig, which had previously achieved an overall depth of 35,000 feet the year before, perhaps leading British Petroleum to consider the new well within normal parameters. But the presence of methane deposits changed the equation. Methane gas under extreme pressure combines with cold water to form dangerous methane hydrate, a flammable, icy sludge that irretrievably fouls oil lines and can congeal into a solid block that plugs them shut. Any sudden decrease of temperature or pressure allows the methane to escape as a gas, rapidly expanding to over 100 times its initial volume with the force of an explosion.

Fracking has completely shifted the geopolitical economic landscape, opening up vast reserves and allowing the United States to become a net exporter of oil for the first time in decades. Yet Big Oil ignored fracking's potential even longer—another example of industry-wide groupthink. Fracking, or hydraulic fracturing, forces water (and other, less savory chemicals) into shale beds under high pressure, splitting the formation and allowing trapped gas and oil to flow freely. The current renaissance is largely due to the efforts of one man, George Mitchell, the so-called father of fracking, who demonstrated that the technique could be economically viable.

Methane Hydrate: The Good, the Bad, and the Ugly

Despite methane hydrate's decidedly negative effects on oil drilling, researchers from around the globe are exploring its potential as a fuel source. Methane is the primary component of natural gas, and methane hydrate is merely methane in another less tractable form. Although serious exploration has yet to begin, methane hydrate reserves are estimated to be many times greater than the current reserves of natural gas. An economical (and safe) method of extracting it could revolutionize the industry. Transporting methane hydrate ice may also prove less expensive and more reliable than current systems that focus on liquid natural gas.

Around that time, one of our colleagues who had attended the 2006 International Nanotechnology Conference accepted a very prestigious position at Rice University to fill the shoes of the late Dr. Richard Smalley, who had shared the Nobel Prize with Sir Harold Kroto. We began to collaborate with the Rice nanoengineering program. Rice University is a major force in materials and nanotechnology, and it is located in Houston, the center of the energy world. So when we showed off our cool Anhydra surface, everyone in Houston was impressed. That served to begin a conversation about what was actually needed.

Royal Dutch Shell, one of the first companies to really take an interest in Anhydra, had no need for a surface treatment that repelled water—but they were impressed by the technology and its effectiveness. Of particular interest was the fact that we were targeting industrial applications for difficult environments. They asked if we could create alternate surface treatments with different effects: How about a hydrophilic, or oleophobic, or oleophilic, or lipophobic surface?

The science is murky at this scale—it defies our Newtonian intuition. More than anything it requires the ability to experiment, to try something and see what works. But we knew the answer was, yes, we can. Once we had a process in place to create a hydrophobic surface, it turned out we could easily tweak it to create a variety of effects, such as a surface that attracts water (hydrophilic) or that repels oil (oleophobic). So, although Anhydra by itself was really cool, its most important value was to demonstrate that a surface could be engineered to a specific performance. This led to what we call a "platform technology" that could be applied to everything, from aircraft corrosion, yogurt processing equipment, to holding steel casing securely in the ground on the seafloor.

We've since developed multiple projects with Shell. One is an Anhydra-related treatment called DragX that prevents methane hydrates from sticking to a pipe. We did extensive testing at the hydrate lab at the Colorado School of Mines. Our treatment is so effective that you have to view the chart on a log scale to even see the effect—it looks like a mistake. This has the potential to make drilling in the presence of methane much more economical and—more importantly—much safer for the crew and for the environment. Another Anhydra variant, SCIN (Steel-Cement Interfacial Nanobond), is a treatment that significantly improves the bond strength between steel and cement and could possibly have prevented the Macondo blowout by attaching the pipe more securely to the surrounding geology. I discuss SCIN in more detail in chapter 12.

Could Shell, or another oil company, have developed these technologies on their own? Perhaps. They have thousands of talented engineers and scientists on staff. But their narrow perspective makes it unlikely. Yet these are exactly the conditions in which Intellectual Anarchy shines—smart people attacking a problem without preconceptions.

////////

Implications

The ability to manipulate the surface of a material on a nanoscale opens up all kinds of possibilities. We break these new surface concepts into two buckets: Functional Surfaces and Intelligent Surfaces.

Functional Surfaces: Anhydra is more than just a specific surface treatment, it is a platform for creating "designer surfaces," what we also refer to as "functional surfaces"—superhydrophobic or hydrophilic, icephobic, oleophobic, omniphobic—with applications in a variety of industries. Here are some examples:

- Power lines, particularly in the northeastern United States, which accumulate ice and can collapse from the additional weight, could become more affordable, more reliable, and require less maintenance, reducing blackouts and increasing the stability of the power grid.
- Aircraft surfaces, including wings and engines, could be more reliable and operate more safely and efficiently in extreme conditions now unthinkable with today's technology. Deicing would be quicker and less frequent for airlines operating at busy winter airports.
- Impellers in petroleum pipelines could become more reliable and require less maintenance, reducing the cost of oil.
- Fracking, which has radically affected the global petroleum industry, has the potential to be much safer through the use of SCIN, an Anhdyra platform spinoff, which has its own chapter later in the book.
- Implantable medical devices such as coronary stents, with an oleophobic surface treatment, could be more resistant to cholesterol buildup and clogging, saving lives.

Intelligent Surfaces: Such surfaces observe, analyze, notify, and report on the environment or the condition of the base material treated. The range of issues we are working on is broad and open up new vistas for how to consider how surfaces interact with humans and machines. Here are some examples:

- Vigilance surfaces, which track the condition of integrity for small-probability events in high-impact platforms, like aircraft or wells, could help prevent sudden, catastrophic failures. An example: Due to undetected fatigue damage, a 1988 Boeing 737-297 Aloha Airlines Flight 243 suffered explosive decompression in flight, and a large section of the fuselage tore off the aircraft at 24,000 feet. Miraculously, only one person was killed, and the crew managed to land safely at Kahului Airport on Maui. Investigators attributed the incident to metal fatigue and corrosion in what was then a nineteen-year-old plane. What if the components of the aircraft themselves had been able to alert people to their condition before they could fail outright? Upon detecting a significant risk, the intelligent surface notifies the user of increased risk of failure.
- Interactive surfaces, ranging from medical to consumer products that change as a function of the environment or some external controls, and turn off or turn on remotely, based on current conditions and needs. This could result in touchscreen-like readouts and controls applied over complex surfaces, or adding functionality to otherwise inert surfaces, such as in work being done by Carnegie Mellon University and Disney Research to develop a low-cost paint packed with electrodes that could turn an ordinary wall into a touchscreen interface.

//////////

The possibilities presented by the future of interactive, intelligent surfaces are so wide open that people in every industry will have to adopt a childlike sense of wonder to see them all. What if? What if prescription pill bottles could recognize who they belonged to and only opened for that person or their caregiver? What if the ink stamp on an egg told you not only its grade, but could change color to let you know it has probably sat in the refrigerator too long? What if the sheets of a hospital bed were soft, comfortable sensor nets that could replace the battery of intrusive leads taped and clamped to patients? What if a bicycle helmet could give riders "eyes" in the back of their heads?

People will miss out on these "what if" possibilities if their habits and organizational dynamics put up roadblocks to creativity. One of the most common such roadblocks is groupthink. There are many factors that can contribute to groupthink and can leave decision makers vulnerable: deference to an authority, cultural assumptions, or even common agreement that your product or idea is better than a competitor's. Overcoming this bias is particularly difficult because it is nearly impossible to see in the moment.

Groupthink is not simply defeated with Richard Florida's "Creative Class" of the new gentrified city. That doesn't go far enough. Defeating groupthink requires true diversity, including a diversity of cultures, ethnicities, backgrounds, and professions. In turn, that diversity must address affordability and wealth inequality. If only wealthy people live in the gentrified city, then diversity is compromised, and groupthink inevitably takes over. Truly diverse communities can be created or found particularly around university ecosystems, which include historians, philosophers, musicians, and artists.

Peter Theil's recommendation that one skip college and start a company instead is wrongheaded. Given the challenges that face us, we urgently need to impact the future of humans and society with innovation on a scale comparable to taming fire or harnessing electricity. The focus on creating yet another interesting but minimally consequential phone app that might become a Unicorn (regardless of the fact that 90 percent of start-ups fail) comes at a cost to producing real disruptive technology, which is almost always based on fundamental understandings in science or engineering. So, speaking as a person who was the first in my family to graduate college, if you are lucky enough to have this opportunity—go to college! You can still build a Unicorn after you graduate. With your exposure to science, engineering, art, history, English literature, anthropology, Eastern religions, and other amazing subjects, you will be less likely to suffer groupthink. Moreover, education enables you to quickly review what took centuries to develop, and it enables you to stand on the shoulders of giants. It is a fundamental building block. Education matters.

CHAPTER 4

Defeating Functional Fixedness: Avoiding Blind Spots in Science, Engineering, and Design

We all have a blind spot, and it's shaped exactly like us.
—Junot Diaz

In *Games for the Super-Intelligent*, James Fixx relates an anecdote of a high school physics teacher trying to illuminate the concept of atmospheric pressure to his students. He asks them if they can determine the height of a building using a barometer. The correct answer, in the teacher's mind, is to read the barometer on the ground, climb to the top of the building for another reading, then calculate the height via the measured pressure differential. One student, "bright enough to be bored by the obvious answer," comes up with two alternative solutions:

1. Drop the barometer from the top of the building and measure the time until it shatters on the sidewalk below. Calculate the height of the building using the formula for acceleration.
2. Find the owner of the building and offer the barometer in exchange for sharing the height.

"At last report," Fixx writes, "the student was in deep trouble with the school's administrative hierarchy."

The teacher in this anecdote is suffering from, among other things, a cognitive bias called *functional fixedness*—the inability to perceive an object's utility beyond its original purpose. He sees the barometer solely as a device for measuring atmospheric pressure and ignores its other qualities and uses, locking him into a particular solution. The student playfully exploits some of those other qualities to arrive at his alternate solutions. If you've ever used a dime to tighten a flat-head screw or wedged an old shoe under a door as an impromptu doorstop, you've done the same thing.

Snopes.com classifies the barometer problem as an urban legend, but true or not, it's a compelling story that illustrates a number of points. Of particular relevance to this chapter is the fact that the teacher fixates on the barometer as a barometer. I've called that an example of functional fixedness, but it could also be representative of an authoritarian inflexibility that's all too common in modern education. Our education system is configured for efficiency, not effectiveness. Problems with single solutions and no allowance for creativity are easier to pose and easier to grade, and woe to the creative student who tries to buck the system.

The second point is the playfulness of the student's first solution. As Fixx puts it, he is "bored by the obvious answer." While the second of his alternate solutions reads, in isolation, like a lazy student's dodge, the first solution is a clever exploitation of physical principles—just not the ones the teacher had in mind. This sort of playful attitude to problem solving is invaluable for innovation. Psychologist Karl Duncker coined the term "functional fixedness" in 1945 and explored the phenomenon in a series of experiments, the most famous of which is the candle problem: participants are given a candle, a book of matches, and a box of thumbtacks and asked to attach the candle to a cork board in such a way that it won't drip onto the table below when lit. An elegant solution is to empty the box of thumbtacks, tack it to the wall, and place the lit candle in the empty box. Few people hit on that solution, however, mostly attempting to clumsily affix the candle directly to the cork board with thumbtacks.

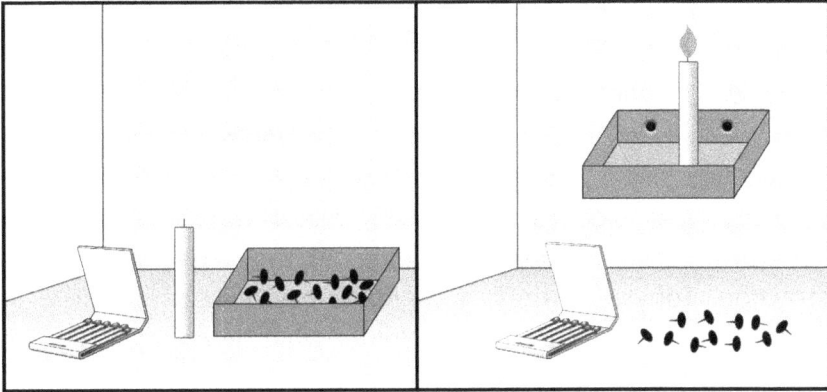

Fig. 4.1. Functional fixedness cripples innovation. Creating self-awareness is essential to addressing this cognitive bias. Duncker's (1945) candle problem is a good example. The subject is asked to attach a candle to the wall, given a box of tacks, candle, and matches, as shown in Panel A. The solution, shown in Panel B, uses the box to hold the candle, not the tacks.

Duncker hypothesized that participants, presented with a box of thumbtacks, saw the box only as a container and not as a possible platform for the candle. This is functional fixedness, the inability to perceive an object's utility beyond its original intention. Later experiments showed that this tendency with regard to specific objects could be enhanced or diminished by cueing the participants with specific language or presentation. If the box is empty to start, for example, people are less likely to see it solely as a container and thus more likely to see its potential as a candle platform. Conversely, if the problem is presented in written form and the word "box" is underlined, participants tend to focus on the label and are less likely to see the matchbox as anything but a container.

A monk sat with his three students. He took out a fan and placed it in front of him, saying, "Without calling it a fan, tell me what this is." The first said, "You couldn't call it a slop-bucket." The master poked him with a stick. The second picked up a fan and fanned himself. He too was

73

*rewarded with the stick. The third opened the fan, laid a
piece of cake on it, and served it to his teacher. The teacher
said, "Eat your cake."*

—traditional Zen koan

Functional fixedness is a specific manifestation of the more
general problem of looking at things at the wrong level of abstraction. Tony McCaffrey, a cognitive psychologist and CTO of
Innovation Accelerator, has developed what he calls the Generic
Parts Technique to train innovators in considering all the potential
features and uses of an object by dropping down a level of abstraction and seeing a candle, for instance, not merely as a source of
light or heat but as an assemblage of parts—wax and string—and
that each one has individual properties. "My research has shown
that people overlook about two-thirds of the types of features
that an object possesses. Not two-thirds of the features, but two-
thirds of the *types* of features. They ignore whole categories that
are not relevant to the object's common use. . . . *This presents an
enormous barrier to coming up with new ideas.*"

At Oceanit, we're very mindful of the traps that labels create, and we're careful to create new language around projects,
especially when they diverge from industry standards. Relying on
stock terms runs the risk of carrying all the baggage of those terms,
the inherent limitations and preconceptions that come along with
them. When we invent a new name for something, we create space
for new possibilities. We do lose out on some initial efficiency. It
takes longer for people to come up to speed with the idea, but
when they do, we know they're on board with *our* idea, and not
whatever idea they've associated with the old term.

One example is our term "anthronoetic artificial intelligence."
Its derivation from "anthro," meaning "human," and "noetic,"
meaning "intellect" is straightforward, but it's an unfamiliar term,
so people don't know what it means. That gives us an opportunity
to explain our approach to strong AI and differentiate it from the
more common big data approach.

People also struggle with visual abstractions. This is a problem that particularly plagues adult beginners learning to draw. They literally must learn to draw what they see instead of what they think they see. They look at a human face, and instead of seeing and drawing the interplay of light and shadow on an irregular surface, they categorize and abstract what they see: a head, two eyes, a nose, and so on. When they turn to their sketch pad, they draw the abstraction. The result is a cartoon. Betty Edwards' book *Drawing on the Right Side of the Brain* is a series of exercises in learning to defeat this process of abstraction.

Experienced artists and designers can fall prey to this trap, too, but if they're clever they can also exploit it to their advantage. If someone complains of heat, you're not doing them any favors by handing them a stiff, irregularly shaped piece of paper. But tell them (or show them) that it's a *fan* and you've solved their problem. When you're trying to bring new users up to speed on a piece of technology, a familiar abstraction can hurry things along.

A few years ago, there was debate raging in the design community on the pros and cons of skeumorphic design—the practice of emulating analog, or real-world devices, in digital interfaces. For instance, a digital radio app might present volume and tuning knobs. The immediate advantage of such a design is that users immediately know how to use it. They're familiar with actual radios and can transfer their existing knowledge to the app. The disadvantage is that it limits the interface to existing modalities and, more important, to modalities that make sense in a physical object. When you look at actual radio apps, they sport features that never would have worked on physical radios—playlists, search functions, and so on.

Ultimately, we just want to keep the possibility of bias in mind. Being aware of functional fixedness is one way to open your mind to innovation. Suppose you were tasked with designing a bicycle helmet but given the "impossible" constraint that it can't touch your head. If you fixate on the label and think of a helmet only as something you wear on your head, you're stuck before you start.

That's functional fixedness. If you think instead of the *function* of a helmet—something that protects your head in the event of an accident—you might develop something completely different, like a collar that inflates to become a personal airbag, triggered by an accelerometer. At Oceanit, we try to create the space to challenge these limiting assumptions.

Functional fixedness can cut in reverse as well—by limiting the results of innovation to the intended application. Many great discoveries were made serendipitously by researchers going after another result entirely. Dr. Spencer Silver, a scientist at 3M, was attempting to create a new, supersticky glue but ended up with a "low-tack" adhesive instead. It took a colleague at 3M, Art Fry, to take advantage of the less-sticky glue to create reusable Post-It Notes, which went on to become a billion-dollar product line for the company. Another example is Viagra, which was originally intended to treat cardiovascular disorders. Closer to home for us at Oceanit is DERT—our revolutionary hemostatic (blood-clotting) agent described in chapter 1. As explained, we were originally trying to create a treated bandage to be applied to the surface of a wound, but our "bandage" had an unfortunate tendency to crumble and fall apart. We could have abandoned the project there, but we discovered that the resulting material, ground up to resemble freeze-dried coffee or dirt, could be packed into the wound to halt bleeding with astonishing speed.

///////////

We have three weapons against functional fixedness, familiar ones at this point: transdisciplinary thinking, getting hands-on, and play.

Transdisciplinary thinking, remember, takes bright, highly educated people and puts them to work on problems outside their field. We've already seen how powerful this technique is at countering expert thinking. It pays similar dividends again here. When someone comes to a new field, they don't have baked-in preconceptions about tools and techniques. They're not immersed in familiar abstractions. They're much more likely to see things as they actually are, rather than as they think they are or how they've been taught they are. They see with the eyes of artists.

Our second weapon is getting hands-on. I explain in chapter 6 how important hands-on interaction is to education and in chapter 7 how important it is to build physical prototypes and actually manipulate objects. Getting hands-on engages our minds in a completely different way. And one of those ways is by helping us engage with an object *directly* rather than with our mental model of it. Remember, Duncker was able to increase experimental subjects' sensitivity to functional fixedness by labeling the physical objects, moving them up the ladder of abstraction.

The third weapon is fostering a sense of play. As the barometer problem demonstrates, creative individuals *enjoy* finding unusual solutions to problems. They thrive on it. We have a group that meets once a week at Oceanit (the S&T group, for Science and Technology). It's like a book club, but instead of books the group brings in strange ideas—an article about a new technology, a novel material, an unusual tool or device—and they'll talk about it, tear it apart, examine it from all angles. There is no end result in mind—it's fun; it's like a game for them.

Recent examples include the following:

- A prototype heat engine, inspired by the lava flow on the Big Island, that converts heat energy directly into electricity
- An AI engine trained in the style of famous artists that can take a selfie and automatically generate a portrait of you in the style of, for instance, Van Gogh
- A general discussion of the feasibility of a space elevator, a device for delivering payloads into near-Earth orbit without rockets, using braided carbon nanotubes

None of these discussions resulted in products or even answers. They just served as conversation starters and idea generators, a fun way to engage at a high level with low stakes—play in its purest form.

Nanite Smart Concrete: Radical New Uses for a Familiar Material

Each material has its specific characteristics which
we must understand if we want to use it.
This is no less true of steel and concrete.
—Ludwig Mies van der Rohe

Concrete is one of the most universal building materials of all time. Bedouins used a form of concrete as early as 6500 BC, and the Romans built their aqueducts and stadiums with it. Concrete is cheap and easy to use, making it ubiquitous—sometimes to the point of oppression. It's not for nothing we speak of "concrete jungles." But even as Soviet-era communists were throwing up cheap, brutal housing projects, a new wave of modernist architects took advantage of concrete's versatility to free them from the classic rectilinear forms of the past.

As with many materials, our use of concrete is limited only by the imagination we apply to it.

Until recently, however, no one's ever looked at concrete as anything but a building material. But what if we could turn it into a sensor? It wouldn't even have to be a very good sensor to be valuable, because concrete is everywhere. It's the second most commonly used building material in the world. Imagine being able to collect information from every freeway, office building, oil well, sidewalk, and parking lot on the planet. The possibilities are unlimited.

The Defense Advanced Research Projects Agency (DARPA) claims to focus on high-risk, high-payoff endeavors. They refer to these projects as "DARPA hard," with the implication that projects that don't fit the profile, by virtue of not being challenging enough, aren't worth their attention. But I've found that DARPA's risk appetite varies up and down with the political climate, and that there are projects that are too risky even for them.

Some of these projects have enormous potential, despite the risk. If we could find a way to mitigate the risk by demonstrating a unique insight or creating a viable prototype, they'd have a chance of getting funded. I began using the term "pre-DARPA hard" for these projects after meeting with a program manager who asked why we hadn't gone to him for funding on a particular disruptive innovation. I told him there was no way he would have considered such a crazy idea before we'd developed the idea sufficiently to take the edge off the risk.

We decided to use our Oceanit Innovation Fund to jumpstart some of these worthwhile-but-too-risky-for-DARPA projects, to push them along the risk curve from "pre-DARPA hard" to merely "DARPA hard." In the last several years we've transitioned nearly 80 percent of these into funded programs, spinouts, or products.

What if we could turn concrete into the world's most ubiquitous sensor and gather data from almost everything? That's the question Dr. V pitched to the Oceanit Innovation Fund: What if we could turn concrete itself into a sensor? What if we could make it talk?

Dr. V had joined us immediately after completing his PhD in mechanical engineering and nanomaterials at the University of Hawaii at Manoa. He's not only a brilliant innovator (and great person) but a fantastic communicator, able to convey complex ideas to a wide variety of audiences, including children. As "Dr. V," he appeared on over 200 episodes of an early morning TV show inspiring Hawaii's *keiki* (kids) about science.

Weird Science with Dr. V

At Oceanit we don't limit our experiments to science in the service of new product development. We also experiment with our business processes and our community engagement. In one such experiment, intended to generate

79

excitement about STEM (Science, Technology, Engineering, and Mathematics) in Hawaii's *keiki* (children, or little ones), our marketing team convinced Dr. V to host a segment on early morning television called *Weird Science with Dr. V.*

Each episode was under five minutes and generally involved blowing something up or setting something on fire. You have to get the kids excited before getting into the scientific explanation. Kids need movement. They need things to happen, *then* you can explain. You can't just present dry theory.

Weird Science with Dr. V aired in 2008 on CBS/NBC–affiliated channels and ran for approximately 200 episodes, garnering a viewership of a few hundred thousand, but Dr. V himself initially resisted the idea of the show. He wasn't keen on "weird science" and didn't want to be seen as anything other than serious. But then one time going through the airport, he was recognized as "Dr. V" by the TSA, and again by a group of kids on the other side of security. He realized he was connecting with people about science and how important that was. That's when he really became Dr. V.

Dr. V had experimented with material science mash-ups earlier that convinced him something like Nanite might be possible. Most recently, he'd been working on improving the heat resistance of foam core concrete building tiles. Regular concrete tiles have been used on roofs since the 1840s as a substitute for ceramic or slate. They're cheap, versatile, and durable, often lasting for a hundred years or more. The addition of a foam core makes the tiles lighter and improves their insulating qualities, but at the cost of making them less resistant to heat. They're not rated for commercial buildings as the foam can burn in a fire.

Dr. V's team at Oceanit created an admix of carbon nanotubes that would turn ordinary clay into "nanoclay." The nanotubes distributed the heat evenly across the tile, preventing hot spots and reducing the risk of burning. This work is still experimental and hasn't been through the certification process, but it did lead directly to the idea of concrete as a sensor. If nanoclay conducted heat, it should be able to conduct electricity as well. (Heat conduction and electrical conduction generally come as a pair, such as in copper, which is an excellent conductor of both.) And if it conducts electricity, it should be possible to use it as an antenna, as a waveguide, and as a sensor.

Waveguides

A waveguide is a device for channeling a wave so it doesn't lose energy as it travels. Waves lose energy as they spread out from their source because the wave front is distributed across an ever-increasing area. In three dimensions, the wave front is an expanding sphere, and the energy falls off in proportion to its surface area, given by $4\pi r2$, where r is the radius of the sphere. This is the Inverse Square Law.

If the wave can be constrained to just two dimensions, the energy of the wave front is concentrated on the perimeter of an expanding *circle*, given by $2\pi r$, and the energy drops off in proportion to the distance from the origin rather than the square of the distance. And if the wave is constrained to one dimension, the wave front becomes just two points receding from the origin, and the wave loses no energy at all.

Consider a pebble dropped in a pond, where the circular wave front rapidly diminishes as it expands, versus a pebble dropped in a narrow channel. The wave is effectively confined to one dimension and loses much less energy. The same principle works for acoustic waves and electromagnetic waves.

High-frequency power lines are an example of waveguides, enabling the transmission of alternating current electricity over thousands of miles.

The Deep Sound Channel (DSC) in the ocean is another example. An inflection point in the temperature and pressure of the ocean water creates a channel that acts as a wave-guide for low-frequency sound waves. The DSC is also called the SOFAR zone (SOund Fixing and Ranging), and submarines use the channel to detect and range enemy vessels. Whales are also known to take advantage of it to transmit their songs across vast distances.

Nanite works in the same way, channeling high-frequency electromagnetic signals with very little loss.

The team developed an admix of specially treated carbon nanotubes specifically tailored to improve the electrical conductivity of cement. They called it Nanite, and it performed better than expected.

The first big surprise was that Nanite's electrical properties changed under pressure. Putting a load on the treated cement created a change in conductivity. We put our first small sample in an Instron material testing machine and gave it a load equivalent to a heavy truck and measured the signal, then unloaded it and measured again. The response was consistent and predictable, meaning we could use it to calculate the weight exactly.

So in addition to conducting electrical signals, every slab of Nanite works as a pressure sensor and a scale. We have a dramatic demo of this in our Technology Petting Zoo, which we show to visitors—a small slab of Nanite you can stand on and use to steer a toy car by shifting your weight. The Nanite is so sensitive that just leaning from side to side is enough to generate an accurate signal.

/////////

Nanite emerged from a deliberate effort on our part to learn more about nanotechnology. As always, we began with a question. Because our initial experience was so limited, that question was necessarily broad—how can nanotech change the world?

Our first attempt at an answer was to do anything we could just to get started. We created our ding-resistant nanotech surfboard (you can read more about it in my discussion of the $10,000 surfboard in chapter 7), then leveraged that experience into nanoclays and finally Nanite. With each project, we explored a different fundamental property of nanotubes: strength for the surfboard, heat conductivity for nanoclay, and electrical conductivity for Nanite.

And with each project, we learned more about how to work with nanoscale materials. We developed techniques for harnessing their potential and scaling up, and we applied those to each successive project. Working on the surfboard, for instance, we had to learn how to mix nanotubes uniformly into resin. It's a hard process, to add something in and have it stay in suspension. The first batches tended to settle out and become unusable. We learned to do it first in water, our breakthrough. We then modified that technique for resin, then clay and cement.

So when I say that Nanite began with Dr. V's pitch to the Oceanit Innovation Fund, that may have been the proximate cause, but it came at the end of an extensive chain of purposeful experimentation and discovery.

If you asked Intel to create a custom sensor, it would be expensive—very expensive. They'd design it on silicon and build it in a billion-dollar fabrication facility, and installing it would require an engineering degree—a typical example of functional fixedness. Nanite was designed to be as simple to use as concrete itself. Unlike most nanotechnology, we designed Nanite so no PhD is required—a high school education is sufficient—because the workforce that mixes cement today has a wide range of education and experience levels. We needed to design our mixing process so the lowest common denominator applicator would always get the right result. Nanite is a simple admix that can be handled by anyone. You pour it into cement as it mixes, and it distributes itself evenly as the concrete cures, ready to talk to you electronically.

This is important because if you can't deploy a technology with an existing workforce, you have to change (replace or train) the workforce, and that increases resistance.

Nanite also has some nice physical properties. It has an indefinite shelf life. You can buy it in bulk, store it, and deploy it as needed. It has a long life in production, and it actually increases the strength of cement. It adds less than 1 percent to the volume of cement and about 10 percent to the cost—another important consideration in price-sensitive industries.

/////////

With all disruptive innovation, there's the issue of moving the technology to market—overcoming the inertia of "the way it's always been done" and educating potential buyers on the benefits. With a product with such wide-ranging potential as Nanite, one of the biggest challenges is identifying the first market, someone for whom the benefits vastly outweigh the perceived risks. Once the technology is out there and in use, others can more easily evaluate whether it makes sense for them. This becomes the classic "Crossing the Chasm" problem for disruptive technology discussed in Geoffrey Moore's book with the same title. Finding the first market—the early adopters—is a nontrivial task. What if we could turn concrete into the world's most ubiquitous sensor and gather data from almost everything? There are potential applications for Nanite in a variety of industries, including transportation, construction, and the oil and gas industry.

Security

Our first project put Nanite to work improving a fenceless security perimeter for a local government facility. Their existing solution relied on an interruptible beam. Anything that crossed the perimeter would break the beam and signal a security detail. But they were plagued with false alarms from pets and wild boar, each of which had to be investigated because they couldn't differentiate them from actual threats. They considered adding cameras for video verification, but the perimeter was so large it would have taken hundreds of cameras to cover it. If they'd had more specific location information to work with, they could get by with a handful of cameras that could quickly zoom in on the intruder, man or beast. That's exactly the solution we proposed. By adding Nanite tiles to the perimeter, they could get notice of exactly which tile was triggered, allowing a camera to automatically zoom in on that location. And as a bonus, because Nanite can detect weight, it could be set to automatically disregard any signal below a certain threshold.

Preventing Bridge Collapse

Our next target application was monitoring bridge health. An estimated 100 bridges collapse annually in the United States, about 4 percent of which result in fatalities. Of those that collapse, roughly half are due to catastrophic events such as earthquakes, floods, and impacts from oversized vehicles. But the other half are "structurally deficient." In other words, their failure could be avoided with preventive maintenance.

Age and usage are the main factors affecting bridge health. The age of a bridge can be measured easily enough with a calendar, but usage data simply doesn't exist. Maintenance is scheduled based on estimates, and then it is often delayed for budgetary reasons, until obvious physical damage appears.

Nanite provides a mechanism to monitor bridge health continuously and schedule preventative maintenance in a timely fashion. The same way a dead lightbulb can reveal a break in electrical wiring, Nanite can reveal hidden fractures in concrete structures. And with Nanite's pressure sensitivity, it's possible to determine if one pillar of a bridge is under increased stress or tension. Note that there's nothing about this approach that's unique to bridges, and the same analysis could be applied to concrete high-rises, including condos and office buildings.

We're in discussions for a program using Nanite to monitor bridge health in seven states across the United States.

Reducing Road Damage from Overweight Trucks

While less calamitous than bridge failures, road damage from overweight trucks is a growing problem. In Hawaii, for instance, our highways operate at 20 percent of their projected life expectancy—just one-fifth of their intended design. A road designed to last for ten years is in bad shape after just one year and severely degraded after two. That means unscheduled maintenance, emergency road repair, budget overruns, and crappy road conditions for 80 percent of a road's life.

Big trucks are responsible for most road damage. A single rig loaded to 80,000 pounds (the legal limit) does as much damage as 5,000 cars. And overloaded trucks are even worse. A truck carrying 90,000 pounds (10,000 over the limit, a 25 percent increase) causes 42 percent more damage than that.

Overloaded trucks are an increasing problem for mainland highways since NAFTA. The legal limits are higher in Mexico and Canada, so we have a problem with overloaded trucks crossing the border. And in Texas and North Dakota, trucks bringing water in for fracking are tearing up the roads. It's estimated that 22 percent of trucks on the road exceed the legal weight limit, but there's no oversight and no good way to track what's going on.

With Nanite, the road itself becomes a sensor, and we're able to measure the weight of vehicles in real time, while the truck is in motion and passing over concrete (e.g. bridges, or concrete measuring points). This solves two problems at once: (1) detecting individual overweight vehicles, and (2) measuring actual road use, as opposed to projected use. Rather than relying on truck drivers to pull into infrequent weigh stations, overweight trucks could be weighed *while in motion* and flagged immediately. And roads could be redesigned to accommodate actual rather than projected loads, based on accurate data about usage.

Currently one company is primarily responsible for weighing vehicles on the road, but the data is locked up. You can't get it, you can't look at it, so no one uses it.

We have a federal pilot project in place to gather weigh-in-motion data on one street in Honolulu. There's no enforcement program in place just yet, but we have a technology champion in the State Department of Transportation who sees the value in this sort of information. This pilot project is just the first step to gathering real information on highway road use in Hawaii and elsewhere. This data has the potential to transform highway safety and longevity. Maintenance costs across the United States vary but average about 16 percent of total disbursements, amounting to $23 billion in 2015. Reducing the cost by 10 percent would save taxpayers $2 billion per year.

Safer Oil Wells

Eighty percent of all oil wells leak because the cementing doesn't hold. Leaks can't be considered accidents or one-off events, as they occur so frequently—if you drill a well, expect it to leak. The industry is aware of this and simply budgets for it. Most leaks are minor, resulting in the venting of hard-to-contain natural gas, but more severe leaks, like the Deepwater Horizon failure, can cause environmental disasters.

After a well is drilled, a casing pipe is extended downhole to keep the sides from collapsing in. Cement is pumped into the gap between the casing pipe and the surrounding rock. The challenge is twofold. First, the cement doesn't always form a secure bond between the steel pipe and the rock, and second, it's not always easy to determine where the cement actually goes. Is it filling the gap as intended? Or has it found its way into a nearby crevice? We're working with oil companies to improve the bond strength between the steel pipe and the cement with our SCIN (Steel-Cement Interfacial Nanobond) product, described in chapter 12. Using Nanite in place of traditional concrete would address the second challenge. By analyzing a signal sent through the Nanite, it would be possible to determine (1) if the cement has gone where it's supposed to go, (2) if there are any unfilled gaps remaining, (3) if there are any leaks or structural damage, and (4) the load on the annulus—the interface between the ground surface, the pipeline, and the ground.

While working on this technology with the major oil companies (more on this in a moment), one of them took a chance with us to demonstrate the basic science on a 40-foot test hole in Hawaii. We're now in the process of developing it for onshore and offshore (e.g., the Gulf of Mexico) applications.

Given all these possible applications for Nanite, we were pretty excited about its potential. Although we now have pilot programs in a variety of industries, we were initially rebuffed by companies that should have been interested. Their resistance was born of a functional-fixedness mindset. They couldn't get past the idea of concrete as a building material only. We met with the head of innovation of a Fortune 100 company who told us, "It's just cement." We explained how Nanite transformed ordinary cement into a waveguide and a sensor, but he had zero curiosity and zero interest. And he was the head of innovation! Likewise, the first company in the oil industry we spoke to—the leading cementing company in the oil patch—laughed at us. They'd locked up their share of the market and had no incentive to change.

Ordinary concrete is so familiar that it's easy to underestimate Nanite's potential. Everyone knows concrete, so they think they understand it. But with Nanite, we're redefining concrete. It's not what it appears to be. That's the trap of functional fixedness. It's hard to break that mindset—and harder still for experts.

It's early days yet for Nanite. We're still identifying the best applications and partners to go forward. And because Nanite targets infrastructure, it's a slow and gradual process. But the uptake has been gratifying. In 2008, Nanite won the NASA Nano 50 Award, which recognizes the top fifty technologies, innovators, and products that impact nanotechnology and industry. And I've already mentioned several pilot projects that are underway. We have launched a major venture with several big players in the oil industry. They're interested in Nanite's ability to measure the curing of concrete in their wells in real time and to monitor the loads on their structures more accurately and continuously than before, particularly in deep ocean conditions where the geologic load can increase dramatically as they pump cement down. We're working with Aramco, Shell, and Chevron to bring Nanite to their oil wells and gas pipelines.

///////////

Implications

Cities are becoming smarter. From London, Singapore, New York, and Helsinki to Montreal, cities around the world are becoming Smart Cities, urban areas that employ various types of sensors, the Internet of Things (IoT), and artificial intelligence (AI) to manage assets and resources efficiently to produce a safer, more reliable, and healthier environment for people to live, work, and play. The implications of smart construction materials are innumerable, including transportation systems, water supplies, wastewater management, law enforcement, libraries, schools, hospitals, and a variety of community services. These will produce better and more responsive environments at a lower cost.

Nanite enables the Internet of Infrastructure (IoI), the next step in the evolution of the Internet itself beyond the Internet of Things. And, as with the Internet, clever people will discover and generate uses for it beyond what we can imagine now. This Semantic Web mashes up functional with information utilization, producing new opportunities and creating new industries, enabling two broad categories:

1. Infrastructure management, including reducing infrastructure cost and improving functionality. These will include the following:

- Faster commutes—with Nanite-enabled smart roads conveying usage information and coordinating traffic in real time to reduce congestion
- Traffic signal improvements—based on actual cars on the road versus timed signals or crude estimates via induction loops
- Intelligent road signs—that warn of traffic delays (again, based on actual, real-time data)
- Responsive traffic routing—for load, safety, and congestion
- Coordination—with self-driving cars
- More accurate traffic data—for online maps and taxi and car-service dispatch
- Ability to tax vehicles more fairly—by miles driven (leveling the playing field with electric vehicles because they don't pay gas taxes)

2. Infrastructure safety and maintenance: the ability to log and dispatch repairs, including the following:

- Fewer potholes—the ability to weigh vehicles in motion creates stricter compliance for overloaded trucks, responsible for most highway damage
- More efficiently manage repair of bridges—continuous monitoring of structural integrity would enable timely maintenance or intervention to avoid collapse from hidden damage
- Improved sustainability—Nanite could track loads, self-manage vibration and noise, monitor fatigue and maintenance needs, could better manage heat, and could track how the building is used, reducing the overall cost of operations and maintenance
- Improved security—detecting anomalies in traffic patterns and vehicle loads could indicate threats

- Earthquake prediction—the highway system could become the largest seismic sensor in the world

The implications of how Nanite transforms one of the most common materials in the world—cement—into an interactive device vividly show the potential of seeing past functional fixedness.

///////////

Functional fixedness is a specific manifestation of the more general problem of seeing things at the wrong level of abstraction. It is essentially a mental block that limits one's ability to use a familiar object or technology in a new and unfamiliar way. However, it's something that is learned, and it's something that can be unlearned. The first step is realizing that we are all at risk of this limitation. One can help to defeat functional fixedness through broad exposure to a range of intellectual and cultural perspectives. Travel, in particular helps one get comfortable with differences and with seeing the world from different perspectives. So can trying different sports—rock climbing, surfing or golfing, or even trying new foods such as Chinese, Italian, or Indian food.

You'll know you're breaking through functional fixedness when you hear yourself saying, "I didn't know that was possible. I didn't know things could be this way."

I witness this all the time when people first visit Hawaii because Hawaii has people from all ethnicities, backgrounds, and cultures: Irish or Kenyan, Korean or Russian, Japanese or Samoan, all are welcomed by Hawaii's host culture. As an island nation, the Hawaiian Kingdom was very welcoming and interested in science and technology. Hawaiians were early astronomers and navigators who sought to integrate technology for the benefit of society. The last king of Hawaii, David Kalakaua, integrated electricity into Iolani Palace before the US White House.

Dr. Paul Romer, 2018 Nobel laureate, discusses the economic consequences of functional fixedness in his paper, "Endogenous

Technological Change" (1990), which expands Kenneth Arrow's (1962) seminal paper introducing the concept. Simply restated, the value of innovation is unlocked when we take finite materials and rearrange them in new ways that produce new applications. For example, if we heat iron ore to roughly 1,200 degrees Celsius, we can make steel, which has many new applications. This is the key underpinning of innovation, resulting in novel ideas that beget new industries and new opportunities that enable economic prosperity.

Endogenous growth is the key to jumpstarting the social construction of technology by increasing the number of people employed in the knowledge economy. Similar to using APIs (Application Programming Interfaces—apps that can communicate with each other) for mash-up apps, there are many interesting and exciting uses found for applications when we encourage people to look at them from a slightly different perspective. This concept sounds simple, and it is, but it is difficult for us as humans to do. We seek out the familiar and comfortable, even with technology, which is evidenced in our tendency to seek skeuomorphic design, which often only slowly evolves into something less familiar and genuinely new.

CHAPTER 5

Defeating Geography:
If You Can Think It, You Can Make It—Anywhere

Geography is destiny.
—Napoleon Bonaparte

"If you're so smart, why are you in Hawaii?"

I get asked that question a lot. The perception is there's little of value here besides beaches, and that Hawaii is nothing more than a vacation destination; certainly not a hot spot for innovation.

Sometimes that attitude is expressed in the form of a statement: "If you're serious, you *wouldn't* be in Hawaii." Technology development in the United States is concentrated in a handful of regions, and we're not just outside the loop, we're remote. Hawaii is 2,500 miles from the continental United States and 4,000 miles from Japan, with very little in between.

Much of this attitude stems from ignorance. There are still people who don't realize that this tropical paradise is part of the United States. Just a few years ago I was carded entering a US military base in Virginia for a meeting. Upon presenting my Hawaii driver's license, I was told I had to be an American citizen. First-time visitors to Hawaii are sometimes surprised to discover that we have running water and adequate supplies of toilet paper, cars, and freeways, and traffic. All this despite the fact that Honolulu is the eleventh largest city in the United States (the City and County of Honolulu, which includes the entire island of Oahu, has a population of nearly a million).

That attitude means we're often not taken seriously. We have to prove ourselves again and again. There are compensating advantages, however. First among them is being outside the groupthink of the existing innovation centers and the culture that brings. Second, the diversity of the population we draw on provides an advantage when it comes to innovative thinking. Third, it's easier to retain talent here, due to quality of life considerations. Fourth is the strength that comes from adversity, requiring us to get surgically focused and very competitive.

Hawaii is an extreme example, but there's a pervasive feeling that technology innovation can only spring from a few chosen cities. That's no longer true. Geography is *not* destiny when it comes to innovation. Or at least it doesn't have to be.

Geographic Determinism

Of course, there was a time when geography *was* everything. Agriculture depends on fertile land, sunshine, and water, and it flourished where those were available. As agriculture scaled up, low-cost labor and access to transportation became the key drivers. Port cities became global hubs, and the appetite for labor spurred mass migrations.

With the advent of the Industrial Revolution, access to energy became a major differentiator—wood and coal, initially, then hydroelectricity and petroleum—but transportation was still important. Waterways were expanded. Fortunes were made as rail augmented and sometimes replaced water transport, tying land and water together, connecting markets and people. Those who controlled the flow of goods controlled the markets. Geography continued to dominate industrial endeavors until the United States passed the Interstate Highway Act in 1956, which connected economies and communities across the country.

There are still geographically determined jobs—for example, the visitor industry in places like Las Vegas, Orlando, and Honolulu, and extractive industries, like coal mining in Kentucky or oil and gas extraction in Texas—but they're becoming fewer and more focused. The tyranny of geography has been slowly undermined by broadly connected energy, transportation, and communication infrastructure. You can get electricity even if you're not in Kentucky, thanks to our national electric grid. You can ship your products to any mainland state via truck or rail, and to Hawaii by sea or air. You can talk to anyone just by picking up a phone.

The Internet is the final nail in the coffin of geographic determinism. It allows almost anyone to reach out to anyone else, anywhere, at any time, and it has created new modes of communication that are still being innovated. With this infrastructure in place, the stage has been set for building a technology business anywhere.

Ingredients of Innovation

Three ingredients are necessary for technology innovation: people, ideas, and capital. Access to capital is a thorny issue that I'll deal with last, but the first two, people and ideas, are available throughout the country.

One of the biggest assets the United States has is its system of over 5,000 universities and colleges. A college education is an enormous enabler, and it is probably the most important thing, next to health care, the country can provide. Great undergraduate and research universities exist in all fifty states, conveying degrees in science and engineering of similar quality everywhere. My daughter attended Trinity College in Hartford, Connecticut, a small, private, liberal arts college, but she studied engineering. They've been offering engineering classes for more than 150 years, and their program is Accreditation Board for Engineering and Technology (ABET) accredited, the minimum standard for engineering programs across the United States. Her classes were taught by professors rather than teaching assistants as at some larger schools, so she got a great engineering education and a great liberal arts education.

I'm sure there are gems like that around the country that just get overlooked. Most university engineering programs are accredited by ABET, meaning they all cover the same fundamentals. Sure, there are rockstar faculty like Richard Feynman, who, in addition to his Nobel-winning breakthroughs in physics, delivered an incredibly popular undergraduate lecture series at CalTech (immortalized as The Feynman Lectures on Physics). But there are great teachers everywhere who can educate and inspire students.

An educated and skilled workforce is critical to building a technology business. Universities and colleges are excellent sources of new talent for the skilled and educated workforce that technology businesses require. And innovation is part of the cultural fabric of engineering programs throughout the United States. Any start-up with access to a local university thus has a plentiful supply of people and ideas. Ninety percent of Oceanit's scientists and engineers are from Hawaii or come via the University of Hawaii.

Of the three ingredients needed for technology innovation, access to capital is typically viewed as the limiting factor. Early stage capital is essentially found in eight regions around the United States: northern California (the Bay Area), southern California, Washington State, Texas, New England, the New York metropolitan area, the Potomac Region, and the Research Triangle in North Carolina. Most of the funding, however, is concentrated in the Bay Area, Los Angeles, Boston, and New York. Outside of these regions the likelihood of attracting venture capital is vanishingly small. For all practical purposes, capital is tied to geography.

But the broad landscape of start-up capital is changing. Marquee venture capital firms like Kleiner Perkins and Sequoia used to get involved more in early stage technology companies, but now that they're managing so much capital, it no longer makes sense. The amount of money they must put to work (often in the tens or hundreds of millions) to maximize their economic benefit in their 2/20 model (2 percent annual fee/20 percent carried interest) means that they are acting more like hedge funds versus helping entrepreneurs launch businesses. More and more, they prefer to get involved later, after the company has inflated to the point that they can invest much larger amounts.

The resulting early stage financing vacuum is being filled by angel investors (individual investors with deep pockets), super angels who invest almost like old-school venture capitalists, and even crowd-funding sources like Kickstarter. While super angels are still concentrated in tech hubs for the most part, angel investors can be found everywhere. And a Kickstarter campaign's reach is as wide as the Internet itself. So, start-up capital is less constrained by geography than ever before.

Defeating Geography

The first step in defeating geography is recognizing that most of the so-called limitations are merely matters of perception.

In Hawaii, one of the major hurdles we have to clear is the perception that we're too far away. But how often do you actually meet face-to-face with the customers or collaborators you do business with every day? As an example, at Oceanit we do work for the City and County of Honolulu. Their offices are just half a mile walk from ours, but we haven't been over there in probably nine months. We established a good working relationship with their staff, and now most things take place by phone and email. They could be on the moon at this point and it wouldn't really make a difference.

This is a lesson I learned from a senior executive at Kaman Aerospace. Founder Charlie Kaman was a major innovator in helicopters, with the first composite rotor and the first electrically powered drone, among other innovations. We've worked with them on a few projects. One of their senior executives would just show up, wherever, whenever he was needed, without acting like it was a big deal. He created the sense that he was always available, that distance wasn't a factor, that geography didn't make him impossible to reach. I'm sure this was the case with everyone he worked with across the world.

That's the behavior we've adopted here. I tell my people to treat the mainland like the "BIG Big Island." We do a lot of business in Washington, D.C., which is five or six time zones away from Hawaii, depending on the time of year. When we're working with DARPA, the NSF, and so on, we just show up as needed; no big deal. We want that relationship to be frictionless at their end. We don't want the perception that we're challenging to work with because we're remote, since it really doesn't matter. So we remove that perception by making ourselves available. Except for frequent flier miles, we are no more remote than Los Angeles, Seattle, or Cincinnati.

Manufacturing

Finally, there's manufacturing. Increasingly, manufacturing is being outsourced across the United States or overseas. However, with great manufacturing partners like Sanmina SCI, Flextronics, and others who manage a global supply chain of materials and components, geography plays a minimal role. Moreover, with 3-D printers, CNC (computer numerical control) machines and other technology becoming available and affordable, geography has lost its hold on technology. In that case it doesn't matter where you live.

When we spun out Hoana Medical—a biomedical company that produces intelligent, sensor-laden hospital beds for improved patient monitoring—we knew we didn't have the capability to manufacture them locally. But we didn't have to. We teamed up with Sanmina SCI, an outsource manufacturing partner in San Jose, similar to Flextronics in Taiwan.

Outsource manufacturers can make full products but also individual components like printed circuit boards (PCBs). We can send them a file—a vector graphic representation of the PCB in a particular format that includes dimensions and part information—and they can print it and populate it with the appropriate electronic components. We can get a prototype in a few days.

They can also drop-ship, meaning they can send the finished product directly to our customers rather than sending them to us, in Hawaii or elsewhere, enabling fast turnaround and just-in-time inventory and finished goods. In the case of Hoana, this was an important consideration for us in choosing Sanmina. Our initial customers were all in the continental United States, and we wanted to minimize shipping time and cost. (Another consideration is that intellectual property theft is rampant in China.)

The Quiet Dive Helmet: Making Stuff Beats Geography

My favorite thing to do—on this planet—is scuba dive.
—Buzz Aldrin

One project where we triumphed over several "geographically advantaged" competitors is the dive helmet we developed for the Office of Naval Research (ONR). They put out a call for proposals and we were selected, along with three others, for the first phase, to deliver a design within six months. The problem they had identified was that the current surface-supplied helmets were so noisy that the divers were all going deaf. All of our competitors were located within 100 miles of Washington, D.C., while we're about 5,000 miles farther away, about as far from the Beltway as you can get.

The US Navy has several thousand "hard hats"—helmets for divers breathing surface-supplied air (in contrast to self-contained breathing systems like scuba). The navy favors hard hat diving for certain difficult and dangerous jobs like salvage, construction, and repair. Because the diver is physically tethered to the surface, it's possible to maintain constant communications contact, though not without difficulty. The interior of a dive helmet is a noisy place—about 110 dB (decibels). That's as loud as a steel mill in operation or a live rock concert and about the threshold of pain for most people. If you've ever had a bucket over your head, you know how it can magnify sound. Now imagine there's a jet of air blasting into the bucket at regular intervals, and add in environmental sound, such as steel tools on ship hulls. Communications volume has to be cranked up to be audible over all that, adding to the din.

The Occupational Safety and Health Administration (OSHA) mandates that workplace noise must be below 90 dB. If that level can't be maintained, hearing protection must be provided. In the event that protection isn't feasible or isn't sufficient to reduce the volume below 90 dB, then OSHA limits the amount of time that workers can be exposed.

Divers can't avail themselves of earplugs—they block the ear canal, preventing them from equalizing pressure—so they must abide by the OSHA limits. At just 92 dB, the limit is six hours. At 100 dB, the limit is two hours. And at 110 dB, the level inside a typical hard hat dive helmet, the limit is a scant thirty minutes. That's not a lot of time to perform an underwater weld or what have you.

ONR called for proposals to safeguard the hearing of the navy's hard hat divers through any means. Ours was the only team that proposed taking on the hardest part, attacking the noise at its source by redesigning the aeroacoustics of the regulator.

The regulator modulates the supply of air to the diver. The dive helmet covers the diver's head completely, like a less aerodynamic version of a full-mask motorcycle helmet, providing impact protection and communications gear in addition to the regulator. Compressed air comes from a tank on the surface through the regulator and into the helmet. When the diver inhales, that lowers the pressure in the helmet. A diaphragm in the regulator is deflected, allowing air in from the high-pressure source.

We bought an $8,000 dive helmet and disassembled it to understand the source of the noise and discovered that the air entering the helmet through a small orifice in the regulator is responsible for most of it. It's like the sound of a gas station air hose for filling your tires—loud and unpleasant.

We'd never worked on a regulator before, but as soon as we took it apart we saw that there weren't a lot of parts and the technology was fairly primitive. One of our engineers, Chris, was excited about aeroacoustics, so he developed a good handle on the physics of noise generation from flowing air. He took the lead on this project and came up with a new design that reduced the noise by shifting the frequency. Humans hear in the range from 20 to 20,000 Hz (hertz). By redesigning the components, Chris shifted the noise to a frequency that humans don't care about.

We rebuilt the helmet and the regulator from the ground up, designing new components with a CAD (computer-aided design) system, producing them locally on our 3-D printer, and ordering other stock parts from an online catalog.

Our first attempt went well, and we managed about 15 dB of noise reduction, which means about 150 times quieter. The starting level was 110 dB, and we managed to take that down to 95 dB. That's an increase in working time from fifteen minutes to four hours, according to OSHA regs. Our final design reduced the noise level by 28 dB, taking the noise level all the way down to 82 dB and well below the OSHA cutoff. At that level, a diver could stay underwater indefinitely. As a bonus, with the noise from the regulator so drastically reduced, the communications volume could be reduced as well.

Theories are one thing, testing is another.
—Brian Schmidt, winner of the 2011 Nobel Prize in Physics

We could have stopped there with a design prototype, but we went the extra step and live-tested it with a diver in the water. We took the prototype to grandma's pool. Chris is a certified recreational diver, and he figured he could hold his breath for a long time if it didn't work, but he brought along a friend, also certified. The pool is only eight feet deep, but any time you're dealing with something that affects your ability to breathe, that's scary, it pays to be safe. He tried the helmet out and confirmed that it worked, not only providing breathable air but at a greatly reduced volume.

At the end of six months, we traveled back to Washington to meet with the ONR with our rebuilt helmet. Of the four companies awarded the first phase, we were the only ones that produced a prototype. Our geographically advantaged competitors arrived with PowerPoint presentations. We arrived with a helmet—a 3-D implementation of our proposed design that had been tested in the water. That put us much higher on the technology readiness level, much more than what you'd expect out of a six-month program.

We'd been outsiders in a literal sense, the only company not on the East Coast. Hawaii isn't known as a manufacturing or industrial hub. There's a perception that we're limited in what we can make or produce. Sometimes it can be difficult to get people to take us seriously. But methods, materials, and tools are changing. Geography isn't the limitation it once was. Across the United States, companies are emailing their design documents to manufacturers around the world. We can do that as easily as they can.

ONR tested our helmet with a navy diver in Washington, D.C., right there in the meeting. There was a guy on board that had been a navy hard hat diver for a number of years. He tried it, and he was impressed. The navy was so pleased with our helmet that they used it as "eye candy" at all their showcase community events, such as Fleet Week in New York City and Portland.

We still have some things to sort out. We have to test breathing resistance at different depths. For example, does our regulator design impose an additional burden on the diver at depth? If it takes too much lung force to deflect the diaphragm in the regulator, that's tiring. Even a fraction of a PSI makes a huge difference. Just 1/360th of an atmosphere, in addition to normal breathing, is too much. Nevertheless, with a few tweaks, we produced a new approach in a field that was considered unchangeable.

Commercial companies that manufacture dive equipment have multimillion-dollar facilities and equipment to test this stuff, including a mechanical lung. We put together our own system for $5,000 to give us breathing resistance information at multiple depths. Our inexpensive setup was sufficient to test our prototype and allowed us to iterate quickly on the design.

We're delivering dozens of these quiet air dive helmet systems to the navy for further testing by navy divers. In the meantime, we're looking at adapting the technology for scuba, fire safety equipment, and more.

The best way to observe a fish is to become a fish.
—Jacques Cousteau

There's an ironic disconnect between the way scuba appears in nature documentaries—the divers gliding serenely through kelp forests and schools of fish—and the actual experience, which is surprisingly loud. Although scuba divers don't have helmets like hard hat divers, they still have to contend with regulator noise.

We hope to dramatically improve the recreational diving experience. We have a prototype scuba regulator, ScubaQ, based on the same technology, which reduces the sound from a menacing, Darth Vader–like rasp to a barely discernible whisper.

We've also taken the lessons we've learned and applied them to related technologies, like the BlastNinja, a quieter blast nozzle, and a redesigned face mask for fighter pilots.

BlastNinja

The lessons we learned on Q-Dive rolled right into work we did on another noise-reduction project, a blast nozzle that's so quiet we call it BlastNinja. Our aeroacoustic skills are taking industrial blasting to a new level, quietly.

Blast nozzles are used to fire an abrasive, such as steel shot or sand, at a painted steel surface to strip it for cleaning and repainting. They are, as you can imagine, extremely loud, and operators have to wear earmuffs and protective suits and can only work a few hours at a time without violating OSHA guidelines. They are also very inefficient. A lot of the energy that should go into propelling the abrasive is instead converted to sound. But no one ever really looked at the fundamental science before we did. We redesigned the nozzle to conserve energy, iterating various designs to provide quiet blasting while still propelling the abrasive. Our final design maintains the effectiveness of the original but reduces the noise by 30 to 40 dB.

Fighter Pilot Oxygen Masks

Based on our success with their dive helmet, the ONR asked us to take a look at the design of their fighter pilot face masks. These supply oxygen to the pilot at altitude. A multipurpose fighter like the F/A-18 Hornet has an operational ceiling of 50,000 feet, nearly twice the height of Mt. Everest. Navy regs require the use of oxygen, but the masks are uncomfortable, so pilots don't like to wear them. Many have chronic facial pain that they've learned to accept as just part of the job. (The next time you meet a fighter pilot, note the thickening of the bridge of the nose from the pressure of the mask; they take on the look of professional boxers.) As a result, pilots will leave their masks hanging from the straps as long as possible and only put them in place when absolutely required. But with a climb rate of 250 meters per second, a Hornet can go from a safe altitude to the death zone in a short moment.

Our goal was to increase compliance (wearing a mask at lower altitudes) by making the mask more comfortable. We leveraged our experience with 3-D fabrication on the Q-Dive helmet to create custom, form-fitting masks for individual pilots, based on a 3-D facial scan. We can scan someone's face and build a comfortable, custom mask that fits in place of the old mask.

Since the mask is made of standard, commodity material, we're essentially producing a "custom commodity," built and fit to the individual's unique face, and manufactured "just in time" for pennies.

While "location, location, location" may still hold true for real estate, for innovation more factors are at play, including diversity, environment, and culture. Where to locate becomes a strategic choice rather than a foregone conclusion. You have to weigh the advantages and disadvantages. At Oceanit we value our freedom from the groupthink of the Beltway and Silicon Valley more than the slight extra effort we have to make to stay in touch. For innovation, for us, the tradeoff is clear.

//////////

Implications

Most industrial systems and tools used today were developed before the turn of the twentieth century when worker safety was typically subordinated to performance and productivity. However, changes in methods and materials have created new opportunities where safety and performance can economically coexist, even if this revolution hasn't fully taken hold. Unfortunately, many of these limitations have since become entrenched in current practice out of inertia—"it's always been that way"— along with the idea that a certain level of permanent physical impairment comes with the job. Indeed, the danger of certain occupations is part of their appeal! This is particularly true in the military (see "Sound Guard: Defending against Hearing Loss" in chapter 9). And most of the hard hat undersea divers we spoke with assumed that worker-related hearing loss comes with the job, because "this is just the way it is." Q-Dive challenges these assumptions. Along with technologies like BlastNinja and Sound Guard, Q-Dive creates a new expectation that job-related disability is something to be managed with innovation rather than accepted as the cost of doing business.

We are at the beginning of an era where the factory is no longer a distant facility at the end of a railroad line or sea-lane; instead it is an appliance in our homes, garages, and workplaces. Here are just some of the ways this could redefine a wide range of industries:

1. Manufacturing is evolving, and it won't turn back:

    •   Filling up a shelf of parts costs money and will become extinct. Just-in-time manufacturing with CNC (computer numerical control) machines, or other broadly used "subtractive manufacturing" tools that have been used for over a decade, has already adjusted the amount of capital allocated to inventory and its associated operational overhead and has shifted the management of the supply chain. CNC machines are affordable and easy to use, and can be located just about anywhere; we can start with a block of aluminum, and a computer can remove material with small drills, augers, and so on to shape what we want with great precision.

- 3-D printing, referred to as "additive manufacturing"— where we add a layer of material at a time—now uses a broad range of materials, increasing in scope and capability almost daily. 3-D printers that range from tabletop to house size can now configure sophisticated designs. At Oceanit we have developed functional fabrics, bolts of fabric made with a machine with a specific functional focus. It could be as simple as a shirt or as complex as a lithium battery separator or a hydrogen fuel cell membrane. We've also developed 3-D printers to print nanocompounds, thermal plastics, wood particles, metals, flexible materials, stiff materials, and more. 3-D printers can even print tissue for medical use or proteins for use as a source of food.

2. Infinite customizability and the ability to meet the demand for increased specificity, effectiveness, and lower costs and accelerate the need for more one-of-a-kind products and solutions:

- A surgeon can order a scalpel that perfectly fits her hand for a specific operation. Body parts, such as ears, can now be printed, and there is even interest in printing/growing a replacement heart. I've noted elsewhere a former DARPA program manager who is building a drug vending machine, sort of like a soda machine but for drugs. All the mixing goes on inside the machine, producing the final prescription when needed. The idea is to make drugs for disease control widely available from something as simple as a vending machine.

- With telemedicine, it is possible for a surgeon and her patient to be thousands of miles apart, and the needed devices can be printed at the surgery location and advanced robotics enable the surgeon to operate via the Internet. Kaiser Permanente has successfully shifted to a digital model enabling many patient appointments to be conducted online, increasing access and proactive, preventative care and ongoing patient follow-up as well as decreasing costs.

3. The new logistics—managing supplies, just-in-time parts, at the location needed, at the time they are needed:

- If you need a part for your car today, you might go to your neighborhood auto parts company and see if they have the part in stock or can order the right part. 3-D printing changes this: if I have a broken taillight, I could download the files and just print a replacement part. 3-D printing is already occurring in aircraft repair at companies such as GE for its aircraft engines.

- One of the biggest challenges for the military is logistics, which essentially means overcoming geography before even engaging the enemy. Unless the army has supplies, including food, medicine, or equipment, it's not ready "to fight tonight." 3-D printing is already starting to change concepts of supply and inventory, reducing the risk of adversaries targeting critical supply caravans.

4. Customized construction—home-building costs and labor will decrease while providing a more customized and affordable dwelling:

- Aside from printing complete homes, which is here today, 3-D–printed supplies would reduce the cost of construction and improve schedules. For example, if I needed a 90-degree, 2-inch elbow to complete a plumbing installation, I could send someone to the hardware store, while the work crew takes a break. However, in the not too distant future, they should be able to bring to the job site a 3-D printer, as simple to operate as a table saw, and print out the part as needed—just as we can do in our labs at Oceanit when we want to create an experimental prototype of a new product.

///////////

The first step in defeating geography is recognizing that most of the limitations are merely matters of perception. Whereas Jared Diamond in *Guns, Germs, and Steel* argues that in the premodern world, geography and environment were key factors that bestowed superior capabilities to certain societies, in today's world, education and key infrastructure are the factors. In all the ways that matter for the Information Age, we make our own geography, building great universities, providing access to broadband, extending power and transportation where needed. With this basic infrastructure, innovation, product development, manufacturing, and more can occur just about anywhere, reducing geography to more of a matter of choice than a matter of necessity.

A second and more profound implication is the idea that existing and emerging manufacturing technologies totally change the calculus of globalization. So rather than argue for the "return of jobs from China," we should argue for faster development of localized technology and increased technical training to prepare the workforce. You can see these changes everywhere, as they are coming increasingly into focus. Former GE CEO Jeff Immelt "broke his pick" while desperately trying to transform GE into a future version with his "Digital Transformation" drive, only to be frustrated at how difficult it was to change the culture of the 127-year-old firm. Yet, at the end of the day, shifting to software and services was likely the right goal—simply, perhaps, the wrong method for the company at that time. Nevertheless, the implications are daunting, vast, and exciting. It's still early, and we are constantly experimenting with new approaches, ideas, materials, and designs for a multitude of applications.

PART III
Innovate

"Innovate" is a verb and requires action—
make it,
break it,
repeat.

Innovations are 10 percent inspiration and
90 percent perspiration.
—Thomas Edison

CHAPTER 6

Out of Their Safety Zones: Interdisciplinary Education and Transdisciplinary Thinking

I have never let schooling interfere with my education.
—Mark Twain

Education is one of the most important innovations the world has ever produced. The ability of individuals to pass on their hard-won knowledge and experience to another, in a fraction of the time it took to acquire, is an incredible feat of intellectual leverage. Isaac Newton meant exactly this when he said, "If I have seen farther than others, it is because I have stood on the shoulders of giants." Without the contributions to physics, astronomy, and mathematics of those who had gone before him, Newton, genius though he was, could never have assembled the whole of calculus and his theory of gravitation in a single lifetime.

In Newton's age, education past basic reading, writing, and arithmetic, with perhaps a smattering of Latin and Greek, was reserved for a privileged few. Newton was fortunate to attend Trinity College in Cambridge, obtaining a Master of Arts degree—the equivalent of a modern PhD—a rare achievement in those days. Giants were few and far between, but their contributions were enormous.

Today, education has become democratized, available to most, if not all. We mint more than 50,000 new PhDs per year in the

113

United States alone. More people than ever contribute to the growing base of human knowledge, though their contributions, perforce, are proportionately smaller. Today it's perhaps more accurate to say that we stand on a mountain composed of the relatively smaller contributions of a vastly larger number of individuals.

With democratization has come standardization, even industrialization. Education has become shrink-wrapped, subjects packaged in neatly labeled, individual boxes. The process has been streamlined for the convenience of educators and administrators, at the expense of students. In fact, our modern education system was *designed* for industry, intended to pump out thousands upon thousands of semiskilled workers for the explosion of factory jobs in the Industrial Age. We're living now in a postindustrial world where globalization, connectivity, and the sheer speed of information are accelerating the pace of innovation. Traditional education must give way to an interdisciplinary approach that will better prepare students for the hypercompetitive global economy of today, where the kinetics of knowledge creation occur faster and faster.

The Prussian Education System

The US public school system, providing free education to every resident through high school, is a towering achievement. At the university level it's considered one of the best in the world, acting as a magnet that has drawn the most talented students from around the globe. But while most would agree that an educated public is essential to democracy, the roots of the system go back to the decidedly *undemocratic* idea of subjugating a population through discipline and conformity.

Horace Mann, widely considered the founder of the Common School movement, the precursor to our current public school system, emphasized the benefits of the discipline it would bring to students. When he went searching for models to build his Massachusetts school system, he became enchanted by the Volksschule (peoples' school) in Prussia, a former German kingdom. The Volksschule was among the world's first publicly funded education systems.

Key ingredients of the Prussian system included isolating the students in rows and teachers in individual classrooms. Mann chose the Prussian model, with its depersonalized learning and strict hierarchy, because it was the cheapest and easiest way to teach literacy on a large scale. The unfortunate side effects of this system—blunted creativity and passion for learning—are with us today.

The principle goal of education in the schools should be creating men and women who are capable of doing new things, not simply repeating what other generations have done.
—Jean Piaget

Let's define some terms.

Modern education, through the undergraduate level, can be thought of as *multidisciplinary.* Each subject is taught as a discrete unit with very little overlap: you learn algebra from your math teacher, you study the Renaissance with your history teacher. This leads to the false impression that different subjects have very little to do with each other. At the graduate level, students are forced to specialize, and crossing between disciplines is actively discouraged. Stepping outside their narrow domain of expertise is deemed too risky. Returning to our metaphor—when we reach the peak of the mountain, we find ourselves surrounded by further peaks representing other disciplines: biology, chemistry, physics, and so on. We are elevated but isolated. While this strategy may make sense for academia, it's a disaster for innovation.

The real world is messy, with no artificial lines between subject areas. Reality is inherently *interdisciplinary*, drawing on skills from a variety of areas at once. While you can work a textbook ballistics exercise (assuming a perfectly spherical frictionless cannonball of uniform density, etc.) with nothing more than a pencil and paper, building a model rocket and predicting its trajectory forces you to discard a number of simplifying assumptions and draw on multiple disciplines. Students need this sort of hands-on connection to the material—making things, breaking things, fixing things. Only by *doing* can they truly engage with the world.

Interdisciplinary education is not only practical, it drives innovation. In *Borrowing Brilliance*, author David Kord Murray claims that all innovation builds on previous innovation. You can't build something from nothing, and the *something* that ideas are made of is *other ideas*. The farther apart those ideas begin, the more powerful they become when combined. Ideas that borrow from within the same field yield incremental innovation. Truly disruptive innovation arises from combinations between disparate fields.

As Murray says, "Any biologist can find solutions in biology, the creative biologist finds them in astronomy. The creative astronomer finds her ideas in macroeconomics. And the creative economist finds them in a Saturday-night poker game—just ask John Nash, the Nobel Prize-winning mathematician depicted in the movie *A Beautiful Mind*."

A multidisciplinary approach to education merely exposes students to different fields. An interdisciplinary approach encourages them to find connections *between* fields.

By *transdisciplinary*, on the other hand, we mean taking the skills and strategies acquired in mastering one discipline and applying them to a completely different discipline. This is a transfer not of domain knowledge but of a way of thinking.

My mentor, John Craven, at the University of Hawaii at Manoa, exemplified this type of thinking, applying thermodynamics to horticulture to arrive at sweeter strawberries, to pick one small example. Inspired by his work on the Ocean Temporal Energy Conversion project (OTEC)—which generates energy by exploiting the temperature difference between cold, deep ocean water and surface water warmed by the sun to run a heat engine—he theorized that strawberries operated on similar thermodynamic principles, producing energy by *delta-t*, the temperature difference between a warm source and a cold sink. He inferred that delta-t would also drive sugar content, which actually produced strawberries 500 percent sweeter than normal.

John Craven, Renaissance Man

Leonardo da Vinci was the original Renaissance man— artist and author, engineer and inventor, astronomer and anatomist. He was a true giant at a time when the world was a smaller place; we may never see his like again. The sheer accumulation of human knowledge has made it incredibly difficult to become a pioneering expert in more than one field. There's simply too much to learn.

Yet polymaths, experts in multiple disparate domains, still exist. My mentor, John Craven, was one such man. He originally obtained his PhD in mechanics and hydraulics from the University of Iowa, after studying engineering as an undergraduate at Caltech and Cornell, then served as chief scientist on the US Navy's Polaris missile program before moving to Hawaii to head up the OTEC program as the dean of Marine Programs, all while authoring several books, including *The Silent War: The Cold War Battle Beneath the Sea*. He also became a lawyer, passing the bar at age sixty-five, and coauthored the Law of the Sea treaty.

John could move so quickly from subject to subject that he kept everyone around him constantly off balance. People either loved him or dreaded him, depending on how comfortable they were with that.

His influence on Oceanit has been profound.

At Oceanit, we depend on smart, highly educated people. As we've seen, however, too much education can lead to narrow, constrained thought. We prefer to recruit scientists and engineers fresh out of their PhD programs, before they've had a chance to develop too many bad habits, and then put them to work on problems *outside* their area of expertise.

That's right. We seek out talented individuals who have spent literally years becoming experts in an extremely specific niche— and then assign them to work on something else.

This counterintuitive strategy makes little sense for companies pursuing *incremental* innovation. Those already deeply immersed in a subject are the ones best positioned to advance the field to the next generation. They know the pitfalls and the promises of various lines of advancement. But at Oceanit we're not just after incremental innovation. We're seeking disruptive solutions to important problems.

By definition, no one has solved an unsolved problem. We find that experts are too easily dissuaded from truly creative solutions by their own extensive domain knowledge. They already "know" what can't be done.

HeatX: The Most Boring Way to Save the Planet

It is up to us to destroy or save the planet.
We certainly have the ability.
—Craig Venter

Some technology is just cool: levitating trains, self-driving cars, artificial intelligence, quantum computing, and so on. And then there's the dreary, not-so-cool tech that's still essential to modern society: mining, waste management, industrial solvents, and others.

Maintaining the heat exchangers in a power plant to keep them free of biological fouling, while essential to keep the modern world humming, definitely falls into the not-so-cool category.

This is true unless you have an innovative surface treatment that allows you to double the effective output of your plant and cut routine maintenance time by 90 percent, increasing efficiency and radically reducing greenhouse gas emissions.

Reducing greenhouse gases is cool. Saving the planet is cool.

We enjoy approaching companies in unfamiliar industries to see if we have anything to offer them. We'll talk to them about their pain points and try to collaboratively identify opportunities for improvements. We've done this with industries as diverse as health care, oil and gas, the local tourism and hotel industry, and education. Despite this diversity and our lack of extensive prior knowledge, we often find that we *do* have something to offer. This stems in part from our transdisciplinary mindset (which lends us an outsider's perspective) and our empathetic listening (part of the Design Thinking process) but also from our interdisciplinary approach to problem solving.

One of the most important factors in education, which is sadly lacking in most educational institutions, is tackling problems that cut across fields. Disciplines are confined to their individual silos, largely for convenience, and problem solving is similarly constrained, leading to a lack of integrated knowledge and often a lack of curiosity, which ill prepares students for the real world.

Worse, we find that this bad habit, acquired in school, extends into industry itself. Professionals are reluctant to look beyond their own fields for solutions to problems, incorporating just enough outside knowledge to get the job done. In some cases, they may not even recognize a problem as a problem; it's just an industry standard to work around.

Because our approach at Oceanit is inherently interdisciplinary, we can sometimes suggest a solution from an unrelated field. And in cases where it's not quite that easy, we have a larger domain from which to develop a solution.

HeatX emerged from a conversation we had with a power plant energy generation engineer at Hawaiian Electric Company (HECO), which produces power for 1.2 million people, with a generating capacity of 2,000 megawatts.

The engineer we talked to was extremely knowledgeable and was willing to take a chance on something new. He's an example of a technology champion, someone willing to buck the status quo and fight for innovation over inertia. These people are rare but incredibly important for moving disruptive tech into a company.

He identified a significant pain point at HECO as biofouling of their heat exchangers, leading to reduced efficiency and expensive maintenance.

Biofouling occurs when microorganisms adhere to untreated surfaces. Left untreated, they accumulate into a layer of scum, like the plaque that accumulates on your teeth if you don't brush regularly. HECO uses coastal ocean water, which includes micro-organisms, to circulate through their heat exchangers, and they'd foul every thirty to sixty days, requiring extensive maintenance and operation juggling to maintain reliable power, at a cost of $1 million to $2 million per year.

We didn't have a solution in hand, but given our success developing nanoscale surface treatments for the oil and gas industry, we were confident we could deliver some significant results in a short time frame. We also had some surprising results from an experimental treatment on a submersible drone that gave us a place to start.

Chesapeake Bay Submersible Drone

A few years ago, we took a flyer on treating a submersible drone for a friend at PACOM (US Pacific Command). The project went well, and we forgot all about it until they called us up and asked, "What is this?"

Submersible drones, referred to as UUVs (unmanned undersea vehicles) or AUVs (autonomous undersea vehicles) are put to work on a variety of missions, including security, situational awareness, defense, industrial monitoring (e.g., monitoring oil and gas com-panies for environmental compliance), and natural resource moni-toring (e.g., assessing the health of fisheries).

To be effective, UUVs and AUVs need to operate over long times and distances with minimal intervention, refueling via solar panels, or periodically swapping out batteries. Like any marine craft left in water for an extended time, they tend to foul and eventually have operational challenges.

Our friend's drone, however, had stayed clean in the Chesapeake Bay for the previous twelve to fifteen months, an unexpected result that gave us something to think about.

Technology Champions

A recurring theme of this book is that people resist change even when it's beneficial, and disruptive change even more so. When trying to introduce innovation to an organization, it helps to have a technology champion, someone one who will advocate for the change and smooth the path.

Technology champions typically start out as obstructionists. They're often corporate subject matter experts (SMEs) with vast experience and will state up front, "I've been doing this for years, and I've heard it all." They're skeptical of anything new. Overcoming that initial resistance is a challenge, but when an SME is converted, they shift from "blocking" to "advocating." They're rare but invaluable.

Our fresh perspective on problems (cultivated by transdisciplinary and interdisciplinary thinking) leads us to creative solutions. We go into a field sometimes not knowing anything, where our customers know everything—or think they do. But they often don't know anything outside their field, and that can make them skeptical of our solution. Their own narrow expertise limits their perspective.

A technology champion within the company, someone with a broader perspective, can serve as a bridge and a motivating force to overcome the inertia inherent in large organizations.

A heat exchanger is a device for transferring heat from one fluid to another without mixing the fluids. Heat always moves from the hotter liquid to the colder liquid, but the heat exchanger itself may serve as a heater or a cooler depending on which of the fluids we're interested in.

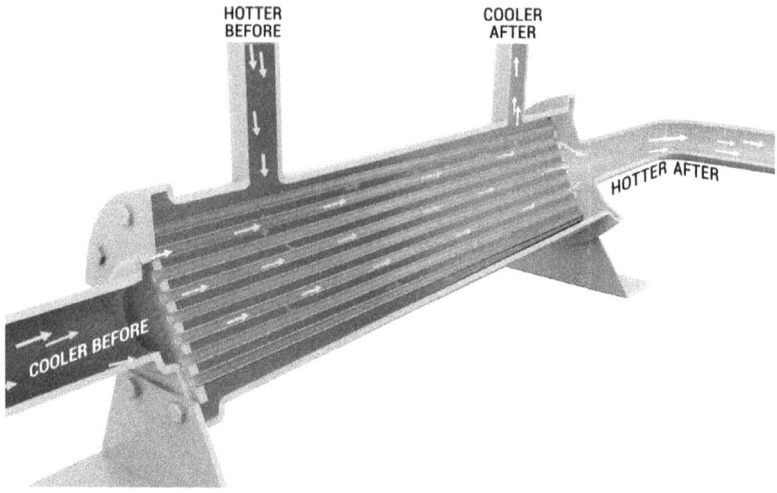

Fig. 6.1. Heat exchangers are used in nearly all modern machines associated with energy, from power plants to desalination plants or processing plants. Heat exchangers essentially "exchange" energy in the form of heat, typically by passing water through metal tubes or plates. Efficient heat exchangers result in less cost, less environmental impact, and smaller physical plants.

In HECO's case, we were looking at two large shell-and-tube heat exchangers that operate as chillers. To generate power, HECO runs purified water in a closed circuit. It's boiled into steam to drive a steam turbine that generates electricity, then the hot steam is passed through a chiller and condensed back into cool water to complete the cycle. This is the red path in figure 6.1 (the shell in a shell-and-tube heat exchanger). Meanwhile, cold seawater runs through the tubes in an open circuit. After absorbing the heat from the steam, the warmed seawater is returned to the ocean.

Biofouling occurs in the tube portion of the heat exchanger because the untreated seawater contains bacteria and other biologicals. The bacteria squirt sticky polysaccharides, like Spider-Man shooting webbing, that allow them to cling to the interior walls of the tubes. As they accumulate, they create an uneven layer that makes it easier for even more bacteria to attach, forming an insulating layer that impedes the efficient transfer of heat. Left untreated, the accumulated biofouling can actually block water flow. If you've ever owned a boat, you know how quickly it can acquire a layer of algae and barnacles that's not only disgusting but can significantly affect the hydrodynamics of the hull. The problem is magnified for the HECO chillers due to the sheer volume of seawater they're moving through the system.

Water Quality

The more microorganisms present in the water, the greater the hazard of biofouling. Coastal seawater, while conveniently within reach, is rich with nutrients from runoff, encouraging more biological activity. The biota from water drawn from Point Panic on the south shore of Oahu would effect a significant change in heat transfer in just thirty days, whereas the cleaner water off the coast of Keahole on the Big Island might last 120 days. And if you were willing to make the effort to pump up the deep water from 4,000 feet off Keahole—this is pure, Antarctic circumpolar water—you could probably go 240 days.

Heat exchangers haven't changed much since they were invented, and the HECO chillers are just standard shell-and-tube heat exchangers on a massive scale with shells 20 feet long and 4 feet across, laced through with 1,500 interior tubes to maximize surface area. Approaches to biofouling are likewise mired in the past, with no significant advancement in decades.

HECO's current strategy is to run two chillers in parallel at less than maximum capacity. They're cleaned on a staggered schedule so that one is still relatively clean as the other is nearing maximum inefficiency due to fouling. At that point, the plant has to throttle down production and take the fouled chiller offline for mainte-nance. Cleaning each chiller can take as long as three weeks and must be repeated every six months or so.

When we first approached HECO, we had nothing in our portfolio that addressed their particular circumstances, but we were confident we could deliver some sort of improvement. The closest things we had were an anticorrosion coating and an antifouling coating. HECO's heat exchangers are titanium, so they didn't need anticorrosion tech (and it didn't contribute to their biofouling issue). And our then current antifouling coating, like most, contained copper as a biocidal agent, which they couldn't use because they ran an open-circuit back into the ocean. Even boat bottom paints that contain copper as an antifouling measure are being phased out due to environmental concerns.

What we did have was an appetite for experimentation and risk and a willingness to look outside conventional industry practices for potential solutions. We also had some recent experience with nanoscale surface treatments we'd developed for the oil and gas industry, in particular Anhydra, a hydrophobic treatment to prevent blockage by methane hydrates in deep-sea oil drilling. We think of Anhydra as a *platform* in that we've been able to tweak it for a variety of applications, from hydrophobic to hydrophilic to oleophobic and so on. Anhydra technology served as the basis for our SCIN surface treatment (Steel-Cement Interfacial Nanobond) for improving the bond between a steel drilling pipe and the cement binding it to the surrounding geology.

For HECO, we took this same technology and tweaked it to create a surface that biologicals would have a hard time adhering to. We didn't know if it would work, or how well, but we hoped it would help a little, at least preventing barnacles.

To test out the new formulation, which we dubbed HeatX, we grabbed a couple of 5-gallon buckets of seawater from the beach. We used a 3 x 3-inch steel "coupon" as a test article. We applied HeatX to one side and left the other side untreated, then immersed the coupon in the bucket of seawater. After a few weeks we found a thick film of bacteria on the untreated side. The treated side was clean.

We knew we were on the right track. HeatX was an extremely thin treatment that strongly adhered to the metal of the heat exchanger, modifying its surface properties, and preventing microorganisms from gaining a toehold in the first place. It didn't insulate, like paint. It didn't contain any biocidal agents like copper or toxins to leech into the environment. And it wouldn't wear away and lose effectiveness like an ablative coating.

The experiment had worked, but we still had to scale up. We try to make progress as quickly as possible because the logistics of scaling up typically take as long as or longer than the initial discovery. We had only a few months until HECO's next maintenance period, the perfect time to deploy HeatX in production. If we missed it, we'd have to wait another six months for the next opportunity.

HECO has a vendor that comes in and maintains their heat exchangers. We knew we'd have to interface with them. That meant getting them onboard with the idea itself—they'd be taking a risk—and dealing with the issues of scaling and training. We couldn't just show up with a bucket of the stuff and hand it over; we needed a partner that we could train to use our technology.

We had tested HeatX only on flat coupons before, so we had to develop a method for applying it to the interior of hundreds of small, round tubes. We bought some sponges at the drugstore, soaked them, and tried pushing them through tubes. We tried filling a tube and tilting it back and forth to distribute the coating. We tried radial spraying inside the tube. We tried adding heat after application to cure the coating, since it didn't dry immediately. Every technique had its pluses and minuses, and we tweaked the formulation to suit the application method.

We started in our lab, then moved to the vendor's shop in Texas, and eventually we got something. By working closely with the vendor, we were able to mitigate the downside to them. There was a learning curve to the application process, but it meant a potential competitive advantage to them if the experiment panned out. We agreed to remain on standby, ready to rush over with an additive if it didn't apply well. If it failed in the field, we'd have to support it.

The vendor applied HeatX to one of HECO's two 43 megawatt chillers with the intention of leaving the second chiller as a control and a backup in case anything went wrong. As it happened, the *untreated* chiller sprung a leak. HECO, carefully monitoring all the way, gradually shifted the entire load to "our" chiller, the one treated with HeatX, for the next four months. It accumulated so little organic fouling that it ran at nearly full efficiency the entire time, effectively doubling its average efficiency.

In fact, HECO decided to skip the next maintenance cycle and let it run. Cleaning the chillers is an expensive proposition, but necessary. Downtime costs run $15,000 per day, so crews run twenty-four hours a day to get back online as quickly as possible. And, of course, while you're offline, you're running at reduced capacity and without a backup.

If both chillers were treated with HeatX, HECO could take either of them offline at any time, knowing they could shift the load to the other chiller, which is capable of running at full capacity. There would effectively be no downtime, no throttled-back production, because you'd have created redundant capacity with no additional capital investment. There's also the savings from the reduced maintenance (we took twenty days of maintenance down to two days, saving about $300,000).

But the really exciting thing is the payoff for the environment. Hawaii is committed to 100 percent clean energy by 2045, but in the meantime HECO burns oil to heat the water that drives the steam turbines and to operate the water pumps for the chillers. An increase in efficiency means less greenhouse gas emissions and a reduced carbon footprint for the plant. HeatX, if used on all HECO plants, would save around 1 million barrels of oil per year, reducing Hawaii's carbon contribution by close to a hundred thousand tons, while reducing the cost of energy. To put this into broader perspective, if we consider Planet Earth as a "burning platform," where power plant carbon contributions from around the world continue increasing global temperatures, we have limited time on the platform before it burns down and the occupants perish. One strategy would be to bring a fire hose to slow down the fire, which would buy more time for the platform occupants to figure out a solution or get rescued. At the current rate of carbon production, the 2-degree Celsius threshold, when the platform burns down, is in about twenty-five years. If power plants around the world did nothing else except improve their heat transfer efficiency with HeatX, the planet would have an additional seven years either to get rescued or to solve the problem. Now, seven years is not much extra time, but now we have thirty-two years to sort things out. Now, imagine more fire hoses showing up: another new technology could slow down the fire spreading on the burning platform, perhaps another seven years. If we keep pushing technology, we can keep the platform from fully burning down and eventually develop a solution where all are saved.

We're currently in talks to treat all of HECO's heat exchangers.

We've also been putting together some numbers for the state of California. They have a mandate to reduce greenhouse gas emissions by 40 percent over the next twelve years. We think we could get them a quarter of the way to their goal just by treating the heat exchangers in their power plants with HeatX. The potential impact is huge, reducing carbon emissions and greenhouse gases for a much more environmentally friendly power strategy.

And down the road are even more exciting potential applications for HeatX. Fouling occurs in a number of environments, whether you're talking about algae or methane hydrates or even cholesterol. We're currently working on a variant of HeatX to prevent clogging of medical stents in the arteries of patients with heart disease.

////////

Implications

The HeatX treatment has broad implications for the bottom lines of many businesses, since nearly all power producing or refining technologies use heat exchangers. In terms of economic impact, HeatX can yield more energy with less cost and less impact. The cost of fouling within a power plant or refinery is approximately equal to 5–10 percent; this includes the cost of running oversized exchangers at maximum load, planned shutdowns that require backup nonrenewable fuel to be burned, and the cost of cleaning maintenance. But by far the biggest cost driver is simply the additional amount of fuel that is required since fouling lowers the heat transfer efficiency of the heat exchanger.

And HeatX is deployable now. Moreover, HeatX could by itself help save the planet from the impacts of climate change. Consider that there are just under 9,000 power plants in the United States and more than 60,000 power plants around the world. If you lock in the 2 percent reduction in CO_2 emissions early by reducing the inefficiencies of fouling, it has an exponential effect on CO_2 emissions going forward. For California alone, that's a reduction of about 4 to 8 million metric tons of CO_2 through fuel savings at refineries and power plants. Even more importantly, it gets a reduction in one of the only sectors—industrial refineries—where emissions actually increased over the past year, a sector where policy pressure has had limited effect. Imagine the possibilities within other industrial factories as well.

To meet the Paris Agreement target of a 26–28 percent reduction from 2005 levels by 2025, the United States will need to reduce energy-related CO_2 emissions by 2.6 percent on average over the next seven years—and faster if declines in other gases do not keep pace. That's more than twice the pace the United States achieved between 2005 and 2017 and significantly faster than any seven-year average in US history.

HeatX could reduce the entire emissions of the United States by 1–2 percent, or around 60 to 120 million tons of CO_2, which could be realized immediately and would nearly get us back on track to meet the Paris Agreement target all by itself, for at least one year. Even nonrenewable energy production becomes greener as a result. Various studies have shown that heat exchanger fouling can contribute up to 1–2.5 percent of all CO_2 emissions in the world.

Another potential HeatX application is in Cloud computing. Cloud storage produces a lot of heat as a result of warehousing massive data, computation, and supportive electronics. Consequently, thermal management is one of the industry's biggest issues and potential costs, requiring major investments in heat management. For an idea of how much heat Cloud storage can produce, consider that the city of Stockholm, Sweden, has been exploring a joint venture between Stockholm Data Parks, city government, and the heating and cooling utility, Fortum Värme, to capture enough heat to supply 2,500 apartments—and, eventually, 10 percent of the entire city's residential heating needs. Similar projects are underway around the world. Another solution to Cloud storage heat that I found especially interesting is Microsoft's idea of moving the Cloud into the ocean. As an example, deep Antarctic circumpolar waters off the coast of Hawaii are very cold, just a few degrees above freezing (2–4 degrees Celsius). It's a perfect place for massive Cloud storage, since Hawaii is already connected with fiber. HeatX would keep this heat transfer at maximum efficiency. Sending photons across the planet in fiber takes very little energy.

HeatX could also produce substantial savings in the area of defense. Heat exchangers and heat transfer devices populate all US military services, but one that seems highly likely to immediately benefit from better heat management is the US Navy. Nearly all ships have heat exchangers. Enabling them to run more efficiently will reduce their annual fuel operating costs, estimated at $4 billion per year. Additional savings in reduced maintenance downtime would also be realized.

///////////

Interdisciplinary education enables incremental innovation, particularly since new ideas are generally found between disciplines and at the edges. Oceanit's transdisciplinary approach, on the other hand, is a deliberate attempt to take skills and strategies acquired in mastering one discipline and apply them to a completely different discipline. This is a transfer not of domain knowledge but of analytical skills. In today's world, information can be obtained from simple and easily available search tools on the Internet. To enable persistent disruptive innovation, the real value of education is the ability to think and to understand how a theory in one field can shed new light on a completely different field, particularly when applied to totally new and unforeseen problems and problems for which a single or interdisciplinary approach has yet to find a solution. Although contrary to how university education is generally delivered today, transdisciplinary thinking is the skill needed for the future. It is the only way we are going to solve our biggest and toughest problems.

CHAPTER 7

Making Stuff:
Hands-On Innovation

It is better to create than to learn.
Creating is the essence of life.
—Julius Caesar

When we're young we learn by doing, by making things. A typical kindergarten class looks more like an arts and crafts workshop than a traditional educational environment—children everywhere with paint on their fingers, surrounded by paste and construction paper. They may look more like they're playing than learning anything substantial, but they are learning. By mixing paints they're learning, intuitively, about the reflective properties of light. By sculpting Play-Doh they're learning about stability and gravity and tensile strength—not by those names, of course, but in a deep and intimate way that means more to them than a graduate student confronted with a force vector diagram.

Unfortunately, this "making stuff" mindset is largely stamped out by high school in the name of efficiency and classroom control. A contributing factor is the class distinction attached to making as opposed to thinking. Making, building, and manufacturing are considered low-class occupations, blue-collar jobs. This same attitude extends even into college, where engineering students are excluded from machine shops and told to hand their designs over to staff who will run the machines for them, depriving them of valuable opportunities to learn by doing. There's great value to making stuff and to the vocational schools that teach those skills.

Machine Shops

I grew up as part of the working class. Most of us didn't have much education; we worked with our hands. Some working-class folks believed that if you "made it," you didn't have to get your hands dirty anymore, drawing a distinction between white-collar office workers and "grease monkeys" in the garage. That's a distinction that may have made sense during the Industrial Revolution, when mass production and specialization became the norm, but in today's rapidly evolving conditions, organizations need to integrate the knowledge and skills of workers at every level.

Today, the distinction between makers and thinkers makes even less sense. Machine shops are not just drill presses and lathes; they include increasingly sophisticated, computer-controlled devices, such as 3-D printers, computer numerical control machines, laser etchers, engravers, mills, and saws.

Safety is always a concern anytime one is around industrial equipment, but too often it's been a pretext for excluding people from using this equipment. Access to a machine shop is *critical* to a good science or engineering education and valuable to most other students. I've seen successful examples where well-endowed shops are reserved for science and engineering students on campus for certain hours, then open to the entire campus after hours, all done with safety and training in mind.

When my daughter was very young, she struggled with math. I brought math out of the realm of the abstract by getting a bunch of cans off the shelf and getting her to explore the concepts physically. (My first hurdle was that, while boys in this situation will eagerly reach out for the cans, girls don't. That's a hot-button topic for another time.) I urged her to pick them up, hold them and move them to get a feeling for math. We played games with colors and volumetric planning. Education makes the most sense most when it's a multidimensional experience. And it worked. By physically experiencing math, she was able to understand it at an intuitive level, connecting her physical understanding to abstract concepts. Today she has a PhD in engineering.

So we embrace this notion of making stuff at Oceanit, of getting our hands dirty, of exploring with all our senses, as well as our minds. We find it's critical to creating the conditions for the innovation lightning to strike, such as rapid ideation, prototyping, and other capacities.

By "making stuff" (the term is deliberately general), I mean a range of activities at various scales, all the way from hands-on "doodling" with paper, cardboard, and tape to new materials, 3D-printed prototypes, and full-scale metamanufacturing.

The virtues of actually making stuff include the following:

- *Experimentation:* Traditional education focuses on recreating the results of past experiments, creating an expectation of "right or wrong" and a "play it safe" mindset, discouraging discovery, whereas making stuff involves actual experimentation and encouraging discovery and creative risk taking. While reproducing classic experiments can be instructive, like using a Michelson interferometer to measure the speed of light, time must be made available for open experimentation.
- *Helping kinesthetic / visual / auditory learners:* Some people (most people, I would argue) simply think better when they're physically, visually, aurally connected to what they're doing. Some people find they have a dominant style, while others use different styles in different circumstances. There is no right mix, nor are the styles fixed. You can develop your ability in less dominant styles as well as further develop styles that you already use well. Not everyone sees the world the same way.
- *Supporting alternate modes of innovation:* My son was a 3-D art prodigy, a prolific producer of amazing art when most kids are struggling to shape Play-Doh. He won scho-

lastic art competitions at a young age with his creations in ceramic, wood, and metal.

One of his teachers told me that he was trying to get my son to write down his ideas in advance, rather than just launching into production. But my son was uninterested (and impatient). He thought it would slow him down. He had these ideas in his head—they were clear to him, he just wanted to get them out, which he did, just about every week for years. We've encouraged him to continue with his art interests, after he got degrees in physics, economics, and engineering design, as a member of the first d.school (Design Thinking) graduating classes at Stanford.

- *Fostering physical intuition:* This is critical at the edges of science and technology, as innovators often find themselves in situations where engineering leapfrogs science; that is, their ability to make something practical is beyond their scientific understanding of how that breakthrough actually works. This is the peculiar position we find ourselves in frequently when working with nanotechnology, for instance, which defies our classical, Newtonian intuition. When we can't get a handle on how to proceed based on firm scientific principles, it's still possible to make progress through intuitive leaps and guesses, which must then be validated by experiment.

- *Uncovering flaws / holes:* Writing down an idea or explaining it to someone else can help clarify it for yourself, alerting you to weak spots that you've glossed over in your mind and forcing you to get specific; similarly, making a physical model can force you to clarify your ideas with even greater rigor. There's a time and place for this, though. Remember, we're looking for ways to make the impossible possible, not the other way around. We don't want to make a model that proves what we're trying to do can't be done. That's not helpful. We want a model that gives us insight into what can be done and how.

- *Settling the debate:* When you're working on the edges of what's known, it's easy to waste time arguing about what is and isn't possible, with both sides bringing rational, credible arguments to bear. In this case it can be faster and cheaper to *just try it*—build a quick and dirty prototype to answer the question empirically.
- *Demonstrating the physics:* Because disruptive thinking threatens conventional thinking, doubters can be slow to come on board. A working, physical model can help sway them.
- *Allowing early design feedback:* I have more to say about this in the chapter 11—"Deep Science to Human-Centered Design"—but a fundamental part of disruptive innovation is making that innovation accessible and useful to the end user. The sooner in the process you can put a prototype or functional model into the hands of a potential customer to gather real-world feedback, the more easily and effectively you can tailor the evolution of the product to its target market.

Of course, the ultimate way to silence doubters is to take the product all the way to manufacturing, something that's often necessary with truly disruptive ideas.

Why Experimentation Matters

Several years back, I was in a meeting in Washington, D.C., where former Secretary of Defense William Perry was discussing concepts in mobile-base deployment technology to address multiple, simultaneous regional conflicts. We developed an idea around floating Lego-style elements that could float in and assemble for a particular application, then disassemble and float away again. You could build, for an example, a temporary landing strip in the middle of the ocean.

Sometime later, I ran into a DARPA program manager in a D.C. area bar, and we chatted about the floating Lego concept. He said that I must know this guy in Hawaii—he's a genius with concrete.

I thought I knew just about everybody, but I didn't know who he meant. He told me about a reusable floating structure this genius had built out of concrete to support offshore drilling in Alaska. You'd float it into position over a drill site, sink it to the ocean floor for drilling, then bring it back to the surface to float to a new site. It employed a honeycomb configuration—a crazy innovation for the time— and a unique cement chemistry that provided excellent protection for the embedded steel rebar. They'd been using the platform for decades with no corrosion of the reinforcing steel, without ever hauling it out of the ocean for maintenance. Additionally, it withstood enormous stress from winter ice that would normally crush it to powder.

This sounded so much like my floating Lego elements idea, only more advanced and actually put it into practice, that I was surprised I didn't know who had done it, especially as he was based in Hawaii. The DARPA guy told me that he's a member of the National Academy of Engineering: his name is Al Yee.

I was stunned. I *did* know Al. His son, Dr. Eric Yee, was a classmate of mine in college, and I used to borrow Al's surfboard when Eric and I went surfing. He'd look at me as if to say, "Don't you dare ding my board." Later we became friends, and he was a mentor to me.

One time, at lunch, I asked Al why he went into structural engineering. He said he originally started out in electrical engineering, but it was too hard to understand. However, when it came to structures, he could see the answer in his head. Afterward he'd have to do the calculations to explain why his answer was correct.

Early on, Al was frustrated with a particular job. He showed that you could make precast, prestressed concrete piles that would hold up to enormous loads, could be used for construction in challenging environments where the foundation was a problem, and could save on construction costs, too. He took this design to his supervisor but was dismissed. He decided to "just do it" in an open lot to prove the concept. He set up prestressed steel and bought a load of concrete to make these piles—and they worked. This is how the Ala Moana Shopping Center in Honolulu was built. This helped launch his career as an innovator.

Al was a true innovator and a pioneer in prestressed concrete. I was fascinated by his thought process—he could see the answer, then had to explain how he got there. I have the same problem, so I could sympathize. It's not the traditional idea of how engineers are supposed to be.

Paul McCartney and Modes of Innovation

Everyone agrees that Paul McCartney of the Beatles is an accomplished musician. He's probably one of the most well recognized musicians in the world, with sixty gold records, thirty-two number-one singles, and eighteen Grammy awards. I was surprised when he said on a recent *60 Minutes* segment that he couldn't read or write sheet music. From a traditional perspective, he might have been dismissed as not serious or limited in capability earlier in his career, though there's no doubt now of his ability to bring his ideas to life. The results speak for themselves.

It's important to remember that there are multiple modes of innovation. Not everyone thinks the same way. To dismiss, out of hand, someone with a nontraditional process might be to lose out on their creativity.

In engineering graduate school, for example, there are two kinds of PhDs—applied and theoretical. Today, the theoretical approach tends to be very computationally intensive. Students spend enormous amounts of time developing models to analyze problems. They must also be able to defend their results with established theories and knowledge. Generally, they need to publish in refereed journals to get buy-in from the broader academic community, as well as to get their committee members, typically five professors, to sign off. Since it's a debate about what's really happening, you may be able to convince a few, but getting everybody to agree is hard.

For an applied engineering PhD, you need to make something happen—make something work that's never been seen before.

You need to explain it, too. But at the same time, you realize that you may see things where there is no complete theory to explain the result. This is where I focused. The reason was that I could always make things and make things work. The proof is in the results, even if others don't agree on why it occurred. If I have evidence and an explanation, if I have produced something never seen before, then I can graduate.

There are multiple approaches to innovation work—top down from theory or bottom up from practical invention; it's possible to innovate without theoretical knowledge. Results speak for themselves with a loud voice.

Hands-On with Nanotech: Why We Made a $10,000 Surfboard

A lot of us at Oceanit enjoy surfing—one of the perks of being located in Hawaii. The waves and the weather are good year-round. One of our favorite breaks is on the south shore of Oahu, just off Diamond Head, a few miles from our office. Known to most as Browns, we call it Grandmom's because my mother-in-law would cook dinner for us after a quick session on Monday nights.

Grandmom's is a slightly treacherous reef break just a bit up the coast from the more aptly named Graveyards and Suicides. At high tide you can glide over the reef without issue, but when the tide is low a thin film of water can give way to coral that seems to pop out and grab your board. This gets even more interesting when the sun goes down.

We try not to stay out past sunset, but sometimes, inevitably, we do. We find ourselves 300 yards out from shore—typically on the second reef, sometimes the third if it's a monster swell—as the light's leaving the sky. When it gets dark and the wind shifts, sometimes we drift off course and the reef takes a bite out of our boards on the way in.

One of our engineers, Ian W., pitched the idea of developing a ding-resistant surfboard at our internal Innovation Fund pitch meeting. We solicit proposals once or twice a year and typically get ten or twelve ideas, which we narrow down to the most interesting three or four. Ian argued that ding-resistant surfboards would make the world a better place. That's true, but not really on the same scale as problems that excite us. What put Ian's idea over the top is that he wanted to use nanomaterials to strengthen the board.

At the time, we were very excited about the potential of nanotechnology but didn't have a lot of experience with it. Oceanit, as noted earlier, had just finished cohosting the International Conference on Nanotechnology, and we were still buzzing from spending time with the world's experts on nanotechnology, including the late Sir Harold W. Kroto. We knew we wanted to do something with nanotech, but we weren't sure how to get started. I'd been pushing to do something—anything—just to get going. We'd learn along the way.

Buckyballs and Nanotubes

Carbon comes in many forms, the best known being diamonds, charcoal, and the graphite in your pencil lead. The specific arrangement of the carbon atoms within these materials is what gives them such distinct properties. Diamonds are strong because their carbon atoms are arranged in a rigid, three-dimensional lattice—a crystal. In graphite, the carbon atoms are arranged in flat sheets of linked hexagonal rings. These sheets are individually very strong but don't adhere well to each other, which is what gives graphite its slippery quality. And charcoal is mostly amorphous (without structure), its atoms arranged haphazardly, and it is accordingly weak and easily crumbled.

In 1985 a distinctive new class of carbon allotropes was discovered, collectively called fullerenes, that form hollow, three-dimensional structures. In the first of these, C_{60} or buckminsterfullerene, sixty carbon atoms are arranged in hexagons and pentagons that fit together like the leather panels of a soccer ball, forming a rough sphere. Named for R.

Buckminster Fuller, who popularized geodesic domes, they are informally called buckyballs. Their large, hollow interiors mean they could potentially serve as cages or carriers for other substances, such as medicine, fuel, or nanomachines. Other fullerenes include C_{70} and C_{540}, which form even larger structures.

Then there are carbon nanotubes, essentially rolled-up sheets of graphene, with a diameter of about a nanometer. They are so incredibly strong and stiff—more than any other known material—that they can form tubes several centimeters long, a length-to-width ratio of more than 100 million to one. For a sense of the scale, a standard 5/8-inch diameter garden hose of the same proportions would be longer than the coastline of California.

Carbon nanotubes also have incredible electrical, optical, and thermal properties. In some configurations, they are fantastic conductors of electricity (1,000 times better than copper and potentially superconducting, according to some experiments), while in other configurations they are semiconductors and could potentially be used for nanoscale digital devices. They are highly directional in terms of heat transfer, conducting heat along their axis ten times better than copper but insulating better than dirt in the other direction.

Ian's surfboard pitch was the perfect suggestion at the perfect time. He's not only a prolific engineer but a surfer and a shaper with a passion for making boards. Ian actually designs the boards that many of us ride, and he's the guy I go to for repairs when mine gets banged up at Grandmom's.

A traditional surfboard begins with an EPS (expanded polystyrene) foam blank, hand shaped by an expert who can tailor the contours of the board to your riding style and skill. The blank is wrapped in fiberglass mesh, cut to fit, and then drenched in resin, which soaks into the fiberglass and hardens into a protective shell.

Our board began the same way, with a blank shaped by Ian. We kept the fiberglass layer, but we customized the resin with an admix of nanotubes. The theory was that these nanotubes would strengthen the resin in much the same way that fiberglass does in a traditional board but on a much finer scale. This actually required quite a bit of experimentation. We started with carbon nanotubes that we synthesized ourselves, but only in very small quantities. We then bought carbon nanotubes in bulk and learned some valuable lessons about sourcing. Not all carbon nanotubes are of high quality, and a batch we bought from China was essentially soot. Everything we built with them failed in testing.

We eventually hit on a source of high-quality carbon nanotubes. We developed a series of prototype nanoresins, essentially pouring the different configurations into a mold that we could examine and test, evolving our ideas with each generation. After realizing the mixtures we needed for strength made the resin look dark, we switched over to titanium nanotubes, even though they're slightly weaker. The carbon nanotubes produced a coal-black board, which looked cool, but we felt a more traditional looking board would better serve the purpose of illustrating the power of nanotechnology. The titanium nanotubes produced a white color that blended in well with the resin.

The result was a board that had the look and feel of a traditional, high-performance board but was nearly 500 percent more resistant to dings, with no weight penalty.

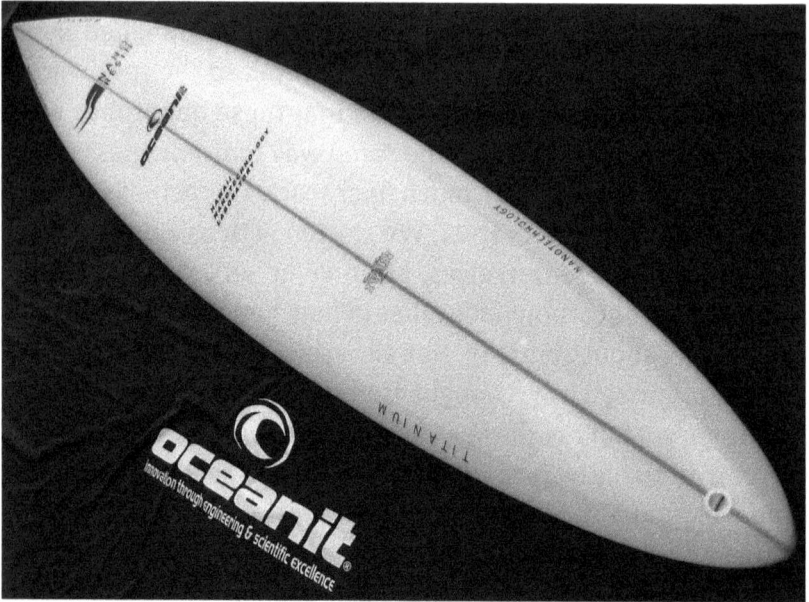

Fig. 7.1. The Nano Surfboard: Titanium nanotubes were mixed with surfboard resin to produce a superstrong finish.

We tested the Nano Surfboard, as we decided to call it, in the lab first. Then we had a local professional surfer take it out for a session in head-high surf for the ultimate trial. He pronounced it perfect. The Nano Surfboard was stronger and sturdier, though it weighed and handled the same as a traditional board. The only downside? It cost $10,000 to make.

Our intention was never to crack the consumer surfboard market, though. We wanted to gain experience with a novel technology with unlimited potential. In that sense, the Nano Surfboard served us beautifully. While we ended up purchasing the nanotubes we used for the Nano Surfboard, our efforts at synthesizing them directly gave us an appreciation for what's possible and experience in manipulating nanomaterials that carried over to other projects like Anhydra, SCIN, and HeatX.

Another key insight was how to mix nanomaterials uniformly. This is an issue when going from the lab to the field or lab to production. We applied these lessons to our smart concrete, Nanite, producing a uniform admixture that can be mixed on-site by an unskilled construction worker and perform perfectly—turning ordinary cement into an intelligent sensor.

We had virtually zero experience with nanotech before developing our surfboard, but we knew it had potential, so we wanted to experiment. And getting hands-on as early as possible is often the key to understanding something new. By actually making a surfboard—though it could have been anything, really—we unlocked the potential of nanotechnology for other projects.

////////

Implications

The implications of nanotechnology are so vast that we don't understand the full range of what it can mean to humans and society. But we can break it into a few big categories, as follows:

1. *Utilization:* The Nano Surfboard is a great example. Early experimentation led to improved strength by mixing titanium nanotubes with the resin. Unless you're a surfer, there's not much to excite you about that. But the same principle, designing new materials with special qualities by titrating nanomaterials into the matrix, can be applied to just about anything, from building materials to clothing. The Nano Surfboard was an important precursor to Nanite (see chapter 4, "Nanite Smart Cement: Radical New Uses for a Familiar Material," where I note that Nanite could also usher in a new era of intelligent, adaptive materials). Other utilizations include the following:
   - *Infrastructure:* Roads, bridges, and other inanimate structures could be information producers. We could, essentially, ask a bridge in the future how it's feeling and get an answer in data form: "Not bad, maybe a

little shaky in suspender cable 327A. And take a closer at my rebar in the west pier, it's feeling too damp."

- *Information systems:* Fixed structures could be enabled to send and receive information, support communication infrastructure, and so on to support a new ITP type network that doesn't rely on energy draining servers. Intelligent paints could change on demand, like a chameleon that changes color to suit its environment.
- *Healing surfaces:* Smart surfaces that sense damage and self-initiate repairs. Oceanit's revolutionary Nanite promises a concrete with sensing abilities. Others such as Henk Jonkers, a microbiologist at the Delft University of Technology in the Netherlands, have developed concrete that heals itself with the introduction of bacteria that produces limestone when exposed to air and moisture. New research is being conducted elsewhere on using gold atoms to facilitate self-healing in metals.

2. *Materials properties:* This is now leading to early experimentation and new product categories. Properties include the following:

    - *Thermal:* This can change the way we dress. Imagine an athletic garment that senses the environment and the athlete's body heat and selectively traps heat or lets it escape to maximize comfort and physiological performance.
    - *Electrical:* "Tech" jackets and hoodies today have pockets and channels for electronic devices and cords, but if made of electrically enabled smart fabric the garment itself could supply power, carry signals, gather data, and serve as a control surface.
    - *Durability:* Outwear for sports and protective gear for hazardous environments could be as tough and resilient as the Nano Surfboard itself.

144

- *Security:* For example, an undershirt suffused with nanofibers, as thin and flexible as cotton, but with the stopping power of a Kevlar bulletproof vest.

///////////

How do we unlock the potential in the implications of nano-technology? Just try something. I recently told one of our young engineers, who is working on undersea broadband, to hurry up and make a crappy prototype so we can move on to a better pro-totype and eventually an amazing technology. He quickly made something that didn't work—but what he reluctantly showed me was mind blowing, transmitting close to a gigabyte per second through water with a laser. We are now preparing to do this in the ocean in about 5,000 feet of water.

We don't always know why things work, particularly when it comes to developing disruptive innovation. But we do know that the key to this is the ability to make stuff, experimental space to try things, make mistakes, and learn. It turns out that innovation—in particular, disruptive innovation—is a messy and multidimen-sional endeavor. If we restrict ourselves to just a few modalities of learning, confining ourselves to whiteboards and computer-aided design (CAD), we don't get the benefit of all our senses. We learn from firing it up, breaking it apart, burning it down. These are all valuable learning opportunities and how real innovation occurs.

Intellectual Anarchy relies on asking good questions that lead to experimentation, which leads to new insights. It tends to be biased toward "Show me," versus "Give me another PowerPoint slide." We debate science and theories, develop models and research plans, but we drive toward a suitable experiment. I find myself oftentimes being delighted to find a surprising result, not what we expected, but from this emerges some amazing insight that leads to an inconceivable or unimaginable possibility. For example, we recently developed the "grammar" of RNA using our anthronoetic artificial intelligence engine that enables the direct engineering of therapies. We've connected this new insight with the ability to engineer and design new antibiotics from scratch—just as we see the world running out of suitable antibiotics for a whole host of common ailments, which have become resistant to the standard antibiotic therapy used across the world.

The "make it, try it, break it, learn" approach has massive implications for developing disruptive innovation. The idea that it's okay to try stuff and fail, as long as we don't fail fatally, is enabling rapid technology deployment. So, having access to tools, creating opportunities for people to make stuff, try things, and experiment is an essential component of this approach to innovation.

It's already possible to make music using computer-based instruments, synthesizers, and audio editing technology. This is transforming music into something that everybody can experiment with, not just musicians. For example, rather than learn to play the violin, I can simulate the violin synthetically and mix this sound with other instruments, all without knowing how to play an instrument. This won't necessarily make me into a composer or a musician, but it can radically change my experience of music.

The overall goal is positive impact for humans and society. With continuously improving technology, making stuff will be available to anyone. It will be this willingness to experiment, to "make stuff"—even crappy prototypes—that will transform society. The generative potential of the materials around us is dependent only on our ability to see the possibilities.

CHAPTER 8

Rise of the Techno Warriors

*I do not think there is any thrill that can go through the
human heart like that felt by the inventor as he sees some
creation of the brain unfolding to success. . . . Such emotions
make a man forget food, sleep, friends, love, everything.*
—Nikola Tesla

At Oceanit we often work on special projects for the US military, either with DARPA (Defense Advanced Research Project Agency) or directly with one of its branches, such as the US Army or Navy. We've developed multiple life-saving technologies as a result of these projects. I've already discussed DERT, the compound we developed to slow bleeding for combat soldiers wounded in the field, and our work on earplugs is discussed in chapter 9, which doesn't sound glamorous but has the potential to mitigate the single most common disability in veterans: hearing loss. Later in this chapter, I look at another such technology: LATCH, which has the potential to revolutionize the way we treat brain bleeds in soldiers with head trauma.

Our work with the military has engendered in us an enormous respect for the men and women who serve, particular those in combat roles. They are warriors, champions willing to lay their life on the line to protect others and defend the ideals of their nation. They are a unique breed: not fearless, but undaunted by fear; not reckless, but willing to embrace risk. They are motivated more

147

by their ideals than such extrinsic factors as pay and recognition, which are all too often in short supply.

We don't face bullets and bombs in the lab, fortunately, so we're spared the necessity for the sort of physical courage that these warriors possess. Yet we find tremendous overlap in qualities between soldiers and those who function well in an environment dedicated to disruptive innovation. A former DARPA program manager called our people "the Special Forces of tech," and we've since adopted the term "Techno Warrior" to describe our ideal researcher. The Techno Warrior embodies four principle ideals. They are (1) self-empowered, (2) intrinsically motivated, (3) passionate champions, and (4) comfortable with risk.

A key characteristic of Techno Warriors is that they are *self-empowered*. They understand their objectives and are proactive and flexible in their approach. They don't require continuous oversight and direction (and may actively resist it). As our friend observed, they are more like Special Forces than traditional military, entrusted with a mission and given the authority to carry it out.

This principle of empowerment is part of a larger strategy we adhere to at Oceanit, drawn from our Strategic Principles that determine our corporate culture (detailed in chapter 9). The idea is that everyone at every level of the organization should be aware of our priorities and authorized to act as necessary to further those priorities.

Given the opportunity, most people want to do the right thing. Given the requisite information—the underlying strategy—most of them can and will. In our experience, this is true more than 95 percent of the time. The occasional misstep is a small price to pay for an engaged, motivated, proactive workforce and *essential* for creative problem solvers. This is a challenge for locked-down, command-and-control style management.

Techno Warriors are *intrinsically motivated*. That is, they are driven more by the challenge of the work itself than extrinsic rewards such as pay raises and promotions. In this regard it matters enormously that the work is worthwhile, which is one reason at Oceanit we like to tackle big problems that matter.

Nikola Tesla

The prolific inventor Nikola Tesla exemplifies this ideal of the intrinsically motivated researcher. Tesla was a pioneer in the early days of electricity, inventing, among other things, the brushless induction motor, championing the use of high-voltage alternating current for power distribution (the system still in use today), and creating a method for *wireless* power transmission when wired power transmission was still in its infancy. Tesla cared more about pushing back the frontiers of knowledge than he ever did about capitalizing on his discoveries, to the extent that others often took credit for his work or took advantage of him financially. He devoted himself to his research, pushing himself relentlessly. He forswore all romantic entanglements as a "distraction" and was never so happy as when he was alone in his lab. Though Tesla did not ultimately reap the financial rewards of his efforts, his accomplishments speak for themselves.

At Oceanit we pay our people well, of course, and they share in the rewards if we spin out a company or do a license deal. But, I always remind myself, that's not why they come here. What we offer them that's even more valuable is the opportunity to tweak the universe, to move the needle of what's possible. In exchange, they bring more value, more passion than we could hope to compensate them for financially. You can pay people to fill a seat, but you can't pay them to bring the magic.

Innovation is challenging work, fraught with pitfalls and obstacles, false turns and dead ends—definitely not for the faint of heart or the easily discouraged. We like to say, "at the tip of the spear there is no roadmap." While it's vital to have people on board who are self-motivated and self-empowered, there is a third quality we look for in our Techno Warriors: *the ability to champion a project they believe in*. We need them to lead with their heart, not just their head.

As I said in the beginning of the book, we encourage choosing a lead researcher from outside the problem domain. This has the key advantage that they aren't bound by their preconceptions. Their enthusiasm for the project remains untarnished by the obvious obstacles ahead. They never say, "It can't be done" and are willing, even eager, to try unconventional, high-risk solutions. After all, the reason a problem is unsolved is typically that the solution is not obvious. A true believer can keep the team pushing forward into the unknown.

Steve Jobs is the classic example of such a champion. At Apple he pushed hard for the Macintosh, threatening the Apple II line then responsible for the bulk of the company's revenue. The first Mac was severely underpowered, closed off, and inflexible compared to the Apple II. It appealed to artists and creative types but was incapable of running spreadsheets and other crucial business software. But Jobs believed the graphical user interface was the future of computing. And he was right. Later he bet the company on the introduction of the iPhone and the choice not to include a hardware keyboard. In retrospect, the success of the iPhone seems inevitable, but at the time of its launch the technical press was less sanguine, predicting that the Blackberry would eat Apple's lunch.

Not all of Jobs' initiatives were as successful (e.g., the Lisa, the Newton, the NeXT computer), but his ability to inspire teams to produce "insanely great" products on impossible timelines was legendary. Bud Tribble, a member of the original Mac development team, coined the phrase "reality distortion field" to describe Jobs' ability to "convince anyone of practically anything."

The biggest risk is not taking any risk. In a world that's changing really quickly, the only strategy that is guaranteed to fail is not taking risks.
—Mark Zuckerberg

Most people are inherently uneasy with risk. The principle of loss aversion—fearing a loss more than you value the equivalent gain—encourages people to play it safe. The Techno Warrior, on the other hand, is *comfortable with uncertainty and risk*.

At Oceanit we counter loss aversion by instilling a mindset of planning to win, rather than trying not to lose. Trying not to lose is essentially conservative, and playing it safe does not lead to breakthroughs. Planning to win is riskier, but the rewards are greater.

We tend to recruit, as already noted, directly out of PhD programs. Doctoral candidates have a demonstrated ability to push back the frontier of knowledge, and they have learned to live with the risk of failure (it's not for nothing they call it a thesis *defense*). Once they become entrenched in academia or a corporate setting, they downshift from disruptive thinking to "application innovation," milking the discovery that launched their career for maximum mileage.

Our challenge at Oceanit is to redefine risk as the new normal. We manage this in part by creating a "fail-safe" environment—acknowledging that failures are an inescapable part of boldness—but even more by hiring the sort of people that thrive on the adrenaline rush that comes from paddling in to a big idea.

Barry Marshall

The annals of science are filled with researchers taking tremendous risks in their quest for knowledge. A fascinating recent example is Barry Marshall, who, along with his partner Robin Warren, won a Nobel Prize for overturning the widely held belief that ulcers are caused by stress and spicy foods. Marshall and Warren believed that a virus was responsible for these ulcers, despite the common wisdom that no virus could survive in the highly acidic environment of the stomach. Still, Marshall and Warren persisted, eventually cultivating *Heliobacter pylori* as a candidate. All that remained was to prove that *H. pylori* actually caused the ulcers.

Marshall *deliberately infected himself* with the virus. Within days, he suffered violent nausea and stomach bleeding. He had ulcers. Marshall and Warren subsequently won a Nobel Prize for their work in this area.

Not everyone is cut out for the role of Techno Warrior—and that's okay, no judgment intended. Not everyone is comfortable with risk. Not every company is pursuing disruptive innovation. There's great value to society in application innovation. Riding that horse is a worthy and legitimate undertaking, requiring consider-able effort and smart execution. And, as we've seen, employing Techno Warriors is not without its own challenges. They don't fit neatly into traditional top-down control structures. They can be challenging to manage. They will fight for what they believe in. But if breakthrough discoveries and inventions are your goal, Techno Warriors are the people you want on your team.

Better Than a Hole in the Head: A Nonsurgical System for Halting Intracranial Bleeding

Trepanation—cutting a hole in the skull—is perhaps the oldest surgical technique on record. We have archeological evidence of skulls with burr holes dating back 5,000 years. Ancient doctors cut or drilled the holes with sharpened rocks to release what they believed were evil spirits responsible for their patients' symptoms. It sounds barbaric, but we still practice trepanation today—to relieve pressure from bleeding inside the skull.

Certain head injuries, blows to the temple in particular, can tear delicate vessels in the brain, leading to intracranial bleeding. Because the skull is a rigid container with a fixed volume, the ex-cess blood has no outlet. The buildup puts pressure on the brain. Too much pressure can result in permanent brain damage or even death.

These bleeds can be difficult to diagnose because there are no obvious, outward symptoms. Actress Natasha Richardson suffered an epidural hematoma (bleeding between the skull and brain) after a minor spill while skiing on a bunny slope. She declined medical assistance at the time of the accident because she looked and felt fine. Some hours later, a headache prompted her to go to the hospital for emergency treatment, but it was too late. By the time she was diagnosed, she had suffered irreversible brain damage and was pronounced brain dead shortly thereafter.

Early bank vaults were constructed of the single toughest material available at the time: steel. But the invention of the acetylene torch—reaching temperatures of 6,000 degrees, well in excess of the kindling point of steel—allowed robbers to cut through the wall or door of the vault in mere minutes. Vault designers responded with layered construction—steel over reinforced concrete, sometimes with an additional layer of copper as a heat sink—so that a single tool was no longer sufficient.

The anatomy of the head works on a similar principle of multilayered defense in depth. Working from the outside in, we have

- Hair—in most cases;
- Scalp—composed of multiple layers of skin and connective tissue;
- Skull—a layer of bone about a quarter of an inch thick composed of dura mater—a tough membrane attached to the skull.

The brain itself floats in a bath of cerebrospinal fluid, protected by all these (and more) layers. Bleeding into the space between the brain and the dura mater creates a subdural hematoma (i.e., below the dura mater). These bleeds are typically fed by veins (at relatively low pressure), so they grow slowly and are thus less urgent to treat—but also harder to diagnose.

Epidural hematomas, on the other hand, occur when the bleeding is between the dura mater and the skull. They are typically fed by high-pressure venous blood, in which case they grow rapidly, tearing the dura mater away from the skull, causing intense pain, and quickly putting pressure on the brain. If left untreated they can cause permanent brain injury or death in a matter of minutes.

Even in cases where the condition is identified, treatment can be a challenge, requiring immediate surgery (trepanation or craniotomy) to cut a hole in the head and relieve the pressure on the brain. For military forces operating hundreds of miles from emergency surgery and routinely subject to concussive blasts likely to cause this type of injury, this can mean the difference between life and death.

We were asked by the US military to look at possible solutions for these types of injuries. Could we develop a simple procedure or medical device appropriate for the battlefield—something compact yet rugged, with a low skill requirement?

One of our research scientists, Peter, thought it might be possible to use a laser to stop the bleed before it develops into a life-threatening hematoma *without opening the skull first.* The ultimate goal would be a handheld, point-and-shoot laser unit that could be deployed in the field. Soldiers with head trauma resulting in intracranial bleeds could be treated on-site in minutes, before permanent brain damage occurred, rather than waiting for transport back to a surgical base.

Lasers have fascinated me since I was a kid. The word "laser" is an acronym for light amplification by stimulated emission of radiation. Unlike ordinary lightbulbs, lasers emit coherent light in a narrow band of frequencies that can be tightly focused. Peter's idea was to tune a laser to a frequency that would be selectively absorbed more by the damaged blood vessel than the skull itself, rendering the skull essentially translucent to the laser.

If you've seen a YouTube video of a laser popping a balloon inside another balloon, the idea is analogous. A blue balloon inside a red balloon can be popped by a ruby laser without popping the red balloon. We see the red balloon as red because it reflects red light and absorbs other colors. Similarly, the blue balloon reflects blue light and absorbs nonblue. The laser, red, passes through the red balloon but is absorbed by the blue balloon, heating and popping it.

Peter took his idea to the physicists, the laser experts, and they all told him it was impossible. The skull would absorb too much energy.

And then there was the obvious problem (obvious even to Peter): How do you aim the laser when you can't even see the bleed?

This is the point at which an ordinary scientist gives up on his clever but ultimately impractical idea. He knows too much, and therefore he doesn't have the drive to persist in the face of the obvious impossibility of the task.

But Peter didn't give up. He decided to experiment. Even if he failed, he'd at least learn something that might point him in a new direction.

Peter's background is in life sciences—he has a PhD in neurology and an MD—but he also has an unusual facility with mechanical devices. He grew up making things for his parents in his home shop, time-saving inventions for tractors, and so on. It's not the sort of thing most people applying for a research position would bother to list on their resume, but it speaks to a uniquely resourceful and problem-solving nature. This practical background plus his deep science education and extraordinary drive make Peter a stellar example of a Techno Warrior.

So, Peter assembled his team. They bought pig heads from a butcher shop in Chinatown, just a few blocks from Oceanit, and rigged a system to infuse them with blood. One refinement we made early on was to create a waveguide in the scalp to fire the laser through. This meant that we would have to find a way to pass the laser only through the skull itself.

With sophisticated software, we created a simulated wound in the pig head, then test-fired the laser. Unfortunately, the physicists were right. The skull absorbed too much energy, setting the skin on fire and shattering the skull. The team burned a lot of pig heads in the beginning. There were a lot of jokes about pork barbecue.

This is the point where even most *extraordinary* researchers would give up. Experiment confirms prediction—end of story. Instead of dwelling on the fact that it was impossible, Peter and his team asked themselves how it *might* be possible. With lots and lots of trial and error—a months-long process that would be reduced to a quick montage in a movie—they discovered a way to manipulate the laser light to pass through the skull.

Along the way, they made a serendipitous discovery. The accuracy of the laser isn't actually that important. The laser coagulates the blood, stopping the bleed without necessarily cauterizing the wound. Remember, this was the second "impossible" condition, and it just fell out for free in the course of experimentation. This is the sort of thing you never discover unless you've embarked on the impossible quest in the first place.

The successful experiments on pig heads have since been confirmed on human cadavers in a major military hospital in Maryland. The result is a device we call *LATCH*, for Laser Actuated Transcranial Hemostasis. "Transcranial" means it operates through the skull—notably, in this case, without opening the skull first. "Hemostasis" is the medical term for halting blood flow, the potentially deadly intracranial bleed.

LATCH is a small, handheld unit, compact enough to be stashed under the seat of a vehicle and simple enough to be used by minimally trained personnel. LATCH has the potential to save thousands of lives per year. More exciting, it points the way toward future technology for minimally or noninvasive surgery with less trauma and quicker recovery times. Variations on the same technique could be used to target specific tissues deep inside the body, unreachable by conventional methods. We may one day be able to look back on surgery with scalpels and bone saws as being as antiquated as leeches and bloodletting.

///////////

Implications

By arresting bleeding inside the skull from blunt trauma, creating a subdural hematoma or intracranial hematoma, LATCH can dramatically reduce the risk of life-threatening or debilitating brain injuries in everyone from athletes to war fighters. LATCH also has interesting implications for minimizing the impact of traditional surgery, too.

It is darkly ironic that in the case of severe internal bleeding, a patient's only hope might be to immediately undergo surgery—essentially sustaining another injury to treat the first injury, with the further risk of bleeding and all the hazards and complications that surgery entail. The LATCH technique could be further perfected to stop bleeding from wounds in other parts of the body, instead of using the traditional hemostatic approaches that require surgical sutures or direct application of physical hemostatic material applied to the tissue.

I can imagine LATCH being further developed to address other traditional therapies that require surgery, such as surgical oncology. Sometimes the trauma associated with cutting the patient open can compound patient risk. I saw this when my father had cancer. He underwent radical surgery to remove his tumor but died anyway. It was never clear whether it was the tumor or the extensive trauma and complications from his surgery that had killed him.

Noninvasive methods like arthroscopic surgery are already becoming widespread, although they still create incisions. Micrographic surgery (also known as Mohs surgery), the technique that removes the cancer and repairs the incision, has become pretty common. A concept worth exploring is treating tumors without cutting open the patient—the LATCH technique. This could lead to outpatient oncology clinics, where one could be scanned for tumors and have them laser ablated (removed) and be sent home the same day without the trauma of surgery, reducing risk from complications and infections and increasing survival rates, as well as reducing costs and improving outcomes. Perhaps with more work, LATCH could even be used for something like skin cancer, specifically basal cell carcinoma, which occurs between the top and bottom skin layers. Today that's treated with surgery that results in a hole punched through the skin, which is later stitched up. Micrographic surgery has become pretty common. But if perfected, laser energy could be transmitted through normal skin and into a tumor, and there would be no need to cut through the skin.

//////////

Peter's persistence in developing LATCH against all odds was a classic Techno Warrior performance. Just as the Samurai have an unwritten code called Bushido, the warrior code, our experience is that Techno Warriors follow a similar unwritten code. The Bushido code of Virtues includes conviction, courage, veracity, honor, and loyalty. Techno Warriors embody five analogous principles:

1. Courage: They are comfortable with risk.
2. Honor: They honor the scientific method, data, and math.
3. Conviction: They are passionate champions.
4. Intrinsic motivation: They realize that money matters, but they cannot be bought.
5. Loyalty: Although out of fashion in the start-up world where "tech mercenaries" are used extensively, loyalty is earned and given. It's also required to face down daunting, seemly impossible challenges.

For a scientist, a warrior code means the fearless pursuit of solutions, pushing the edges of perceived reality to invent the future. Techno Warriors don't follow a rule book, but they embody a set of strong moral and ethical values that underpin the work that they do. For in the realm of innovation, where the work and its implications can be unknown and scary, Techno Warriors must have confidence and conviction in their pursuit and judgment that their work will help humans and society, for the good of people, the environment, and our world—Planet Earth.

CHAPTER 9

Principled Anarchy:
Enabling Emergent Greatness

Culture eats strategy for breakfast.
—Peter Drucker

Most companies are built to manage, which can come at the expense of innovation. Oceanit is built to innovate but is challenging to manage.

While a vertical, command-and-control management style can make sense for some businesses, such as a manufacturing plant, today's knowledge workers increasingly benefit from a flatter, more horizontal style that pushes decision making down the chain and empowers individuals to use their judgment. Even the US military, one of the most rigidly hierarchical organizations on the planet, is increasingly relying on Special Forces teams with more training and more autonomy, who are able to make decisions in the field. For maximizing effectiveness, it's important. For innovation, it's essential. Any business that depends on the intelligence of its people is going to create more value and opportunity by giving them more freedom to exercise that intelligence.

Flatter organizations are less susceptible to what I call "mission-critical core dilution." Start-ups have only enough resources to focus on mission-critical, core activities, such as making a product, and may poorly manage some of their important but noncore business activities—what we refer to as "mission-critical context." At the risk of oversimplification, mission-critical, core thinking tends to emphasize survivability—focusing on essential capabilities, such as getting a paying customer. Alternatively, mature organizations sometimes are managed to the detriment of their core mission or offering; for example, they add staff and resources so they can never be blamed for not doing their job. This risk averse behavior comes with maturity, but sometimes it goes overboard. This is not a simple balance, since context issues are essentially table stakes for business, and if not managed well, they result in getting dinged or punished, or worse; for example, if a commercial airline loses your bags, you are unhappy and complain. However, if they have perfect bag delivery but do poorly with safely transporting passengers, they are in trouble, since their core business is safely transporting paying passengers. Paying, happy passengers are mission critical, core to the business; bags are context. Start-ups tend to ask for forgiveness when they overlook a permit filing, tax filing, or other paper work. Mature businesses become risk averse and never want people to think they are not doing their job, even if they forget what their core business really is or how their core business has evolved.

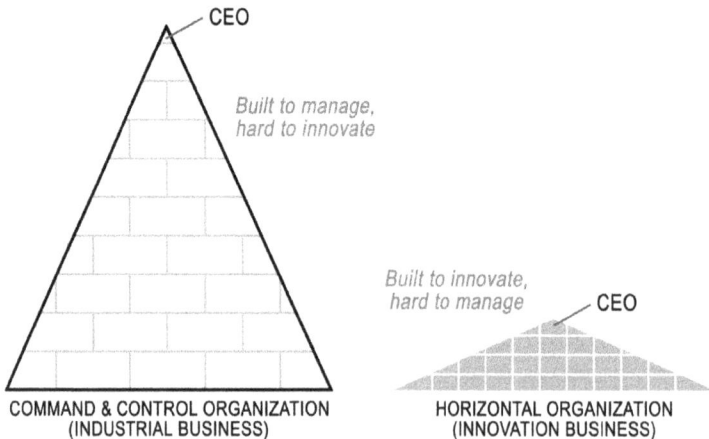

Fig. 9.1. Most industrial organizations are managed in a vertically stratified "command-and-control" framework. Technology / innovation-based companies with a highly educated workforce tend to be flatter, with fewer layers of management.

Strategic Principles

For people to do amazing things, you must create the environment and expectation for greatness. At Oceanit we have a set of Strategic Principles developed over the past thirty years that serve as our high-level strategic concepts and guidelines. We give our employees guidelines, not rules—"What," not "How." This concept is scary for traditional managers. They must sacrifice a certain measure of control. What if you trust someone and they do the wrong thing? But we find that most people, most of the time, will do the right thing, as long as you make it clear what your expectations are. And the benefits outweigh the risks. Working in this fashion creates an environment for "emergent greatness."

It's an enduring frustration of working with computers that they do precisely what you tell them to do, which may not always be what you actually want or need. One of our long-term projects at Oceanit is artificial general intelligence—a machine that is as smart and flexible a thinker as a human being. But we already have *natural* general intelligence in the form of human workers, which too many organizations insist on treating as simple machines, enforcing rigid policies and affording little opportunity for creativity or personal judgment.

We've seen this model work beautifully at Oceanit and other organizations. In particular, at the Stanford School of Design (d.school). We've partnered with them on a number of projects and had the chance to see how they operate. They don't have a lot of rules for doing what they do. They control for two variables—the quality of the students and the quality of the faculty—and believe that great results will follow. And they do. When you put great people together, they find a way to accomplish great things.

Conversely, we've seen the disaster that can come from disallowing judgment, as in the case of the doctor who was dragged off a United Airlines flight. That's a rules-based system. On an overbooked flight, the flight crew didn't have the discretion to offer a sufficient reward to entice a passenger to give up their seat. Reportedly, one passenger even offered to deplane for a few hundred dollars more than what was offered. Instead, the flight crew selected four passengers to deplane at random. When one refused to yield his seat, they called airport security to remove the doctor by force, to the dismay of other passengers who recorded the incident. The high-profile case went viral, and United's stock was hammered to the tune of $250 million. If the airline had a set of strategic principles that, hopefully, included passenger well-being as a primary concern, this never would have happened. The crew would have found another solution.

At Oceanit, our principles are designed to maximize executing our mission. We inform people of the ground rules with enough flexibility to allow them to think. And we hire well-educated people who we trust to make good decisions.

Diversity

We need diversity of thought in the world
to face the new challenges.
—Tim Berners-Lee

One of our most important principles is "embrace diversity," which is another way of saying "encourage curiosity." If you're afraid of what's different, then by definition you're afraid of what's new. That's not healthy when you're looking for innovation. Being open to people means being open to their ideas and their point of view on the world—and vice versa.

Intellectual Anarchy is all about generating solutions to difficult problems by bringing a fresh perspective to bear. The greater the diversity of our research team, the easier it is to manage. Diversity also counteracts groupthink and functional fixedness, two cognitive biases that can impede innovation.

Today, monocultures like Singapore and Japan are struggling with innovation. Innovation hubs, cities like New York, San Francisco, and Honolulu, are more diverse.

In Hawaii, diversity is a natural part of the community. Honolulu is one of the most diverse cities in America. No single ethnicity makes up as much as a third of the population. Nothing exemplifies Hawaii's diversity more than the lineup at a surf break, where people of all races, professions, and backgrounds line up side by side to catch waves: *keiki*, grandmothers, lawyers, high school dropouts, engineers, beach boys, yoginis, amateurs, artists, pros, and pets.

We have an engineer here, Dayan, who grew up in Sri Lanka. His father was a village cultural doctor, which gave him a different perspective on medicine, among other things. When I feel a cold coming on, I reach for NyQuil, but he takes turmeric to boost his immune system. When we were working on a coastal resiliency project together a few years ago and needed to clear some large rocks, my first thought was *tractor*, but his was *elephant*. He's got a PhD in engineering but brings a unique perspective to the work. He thinks way outside the standard Army Corps of Engineers design. We used to propose sophisticated designs that incorporated coastal protection concepts used in Denmark—coastal resiliency is a national security mandate for the Danish—only to have these ideas dismissed or rejected. Today, however, US coastal policy is developing a much more flexible disposition due to climate change.

> *Our brain only allows us to perceive data*
> *that matches the belief system we walk in with.*
> —David Allen

It's easy to discount the value of alternate perspectives because we don't know what we're missing. As Defense Secretary Donald Rumsfeld famously said, "There are known knowns; there are things we know we know. We also know there are known unknowns; that is to say we know there are some things we do not know. But there are also unknown unknowns—the ones we don't know we don't know. . . . It is the latter category that tend to be the difficult ones."

In our work, we are constantly grappling with the unknown unknowns, looking for solutions where it's not clear that a solution even exists. But one of the biases we have as scientists is that we believe that we're objective. We tell ourselves we're just looking at data without biases or preconceptions. But, as the David Allen quote makes clear, even the collection of data is subject to bias. We design our experiment to make certain measurements and exclude others. We dismiss data that doesn't seem relevant or, more damning, fail to perceive it in the first place.

A beautiful example of this is a psychology experiment in selective attention by Daniel Simons and Christopher Chabris in which students were asked to count the number of basketball passes by the players wearing white shirts. If you haven't seen it yet, I strongly encourage you to watch the video linked below before continuing (you can also Google "David Simons selective attention test"): https://youtu.be/vJG698U2Mvo

Did you watch the video? Did you get the correct answer of fifteen passes?

Did you see the gorilla?

If you weren't able to watch the video, it begins with an instructional title card, "Count how many times the players wearing white pass the basketball." The video then shows six college students, three dressed in white and three dressed in black. Each team passes a basketball back and forth among themselves: the team dressed in white passes to each other, as does the team dressed in black. Then a series of additional title cards are presented: "How many passes did you count?" followed by "The correct answer is 15 passes," and concluding with the very surprising, "But did you see the gorilla?!"

The video then "rewinds" to show that, in fact, a man dressed in a gorilla suit enters the floor among the other players, stands full center and beats his chest, then walks off. Most people, in fact, *don't* see the gorilla, to the extent that they have trouble believing at first that it's the same video. How could they have missed something so obvious?

The cross-pollination of disciplines is fundamental
to truly revolutionary advances in our culture.
—Neil deGrasse Tyson

Diversity of background also supports cross-pollination, or interdisciplinary thinking. I've commented a lot already about the benefits of transdisciplinary thinking, one of the more radical aspects of Intellectual Anarchy. Interdisciplinary thinking combines ideas from distinct fields or uses ideas from disciplines in an unrelated area.

For example, we've been developing downhole sensor technology for the drilling industry to get accurate information on what's happening 10,000 feet below the surface. We came up with a DNA-based solution. We send a particular molecule down the pipe and bring it back up. We can tell a great deal about the downhole environment from changes to the molecule: the temperature and pressure that it experienced, the chemicals that it was exposed to, and so on.

In the early twentieth century, coal miners would bring canaries with them into the mines as sentinels to warn of carbon monoxide gas, which is odorless and colorless. Canaries, being more sensitive to carbon monoxide than human beings, would die before the miners were in serious danger, giving them time to evacuate safely.

In a way, our technology is an extension of this (with the caveat that, although biological, our "canaries" are not alive). The miners' canary was binary, presenting a single bit of information: too much gas, yes or no? Ours are more sophisticated. Imagine a canary that changes hue to reflect which specific gas it's been exposed to and changes saturation to reflect the concentration of the gas.

We send our stuff downhole, it reacts with the environment, and we retrieve it and do a bioassay to evaluate. We're combining multiple fields: chemistry, biology, mechanical engineering, geology, and genomics.

Perhaps closer to the canary example, we've trained grasshoppers to change the sound of their chirping when they detect specific chemicals. Through millions of years of evolution, insects have developed sophisticated behavior patterns dedicated to a specific "mission profile." If you understand their triggers, you can hijack their behavior, adjusting their profile (to something more suitable). Instead of designing and building one intelligent, complex, and expensive robot, you can deploy a swarm of inexpensive, individually dumb but cooperative, "biological robots" to accomplish the same task. Ants, for instance, come pre-equipped with compact and efficient chemical sensors and built-in stigmergic (indirect coordination) logic. We borrow a lot from traditional robotics in these projects but a lot more from biology, chemistry, and animal behavior.

> *'Ohana means "to work together" in Hawaiian,*
> *which we treat in translation to mean "family."*
> —Pono Shim

A related principle is to "foster *'ohana.*" We treat each other like family, with mutual respect, caring, and sensitivity. We make space for each other. You don't discard family for any reason. As Lilo and Stitch say, no one gets left behind or forgotten.

That does *not* mean, however, that we don't want people to have an opinion or to express that opinion. Not every idea is worth developing, and the filtering process includes vigorous debate on the merits. But we expect everyone to express that opinion with sensitivity—honestly, but with regard for each other as human beings. The fastest way to guarantee that your people never bring up a crazy idea is if you allow other people in your organization to mock or belittle them when they do. Keep the feedback focused on the idea itself, not the person.

Seek Chaos

> *Invention, it must be humbly admitted,*
> *does not consist in creating out of void, but out of chaos.*
> —Mary Shelley, author of *Frankenstein*

Another strategic principle is to "seek chaos." Most companies avoid chaos or seek to control it. They're all about managing risk. In fact, the vertical management structure of most organizations is an attempt to impose order on an inherently chaotic process. But chaos is fertile ground for opportunity. It's only when there is market chaos that market equilibrium changes. Producers and consumers become mismatched. This mismatch creates the opportunity to intercept the producer-consumer relationship. This is how chaos creates opportunity.

The Space Race created opportunities for people to be great. The launch of Sputnik and the credible threat that the Soviets might control space provided the impetus for the Apollo missions, which put the first man on the Moon. Hundreds of bright, young scientists and engineers were driven to tackle problems that no one knew how to solve, and they rose to the occasion.

Sound Guard: Defending against Hearing Loss

Blindness separates us from things,
but deafness separates us from people.
—Helen Keller

In pursuing disruptive innovation, we find it helpful to engage with "extreme users," those that will drive their equipment and themselves to the ragged edge. We had the honor to spend time with the US Navy Seals. A group of scientists and engineers from our office got together with a small team of Seals to "talk story." There was no agenda, just an opportunity for us to understand how we could help them find an advantage at the intersection of what really matters and what physics will allow.

At this particular gathering, they invited us to fire some of the weapons they rely on every day, including the M4 assault rifle, the Sig Sauer 9 mm handgun, and others. We met in a special, secured facility for a safety briefing that covered the obvious—these were real weapons and we had to respect them—and the not so obvious—we'd need to wear ear protection.

If you've never fired a gun, you might be surprised by how loud they are. We've been conditioned by movies and television to think of gunfire as an exciting sound effect, but it's actually loud enough to cause permanent ear damage. For context, the threshold of pain is about 120 dB, and anything above 140 dB can result in permanent hearing loss. A 9 mm handgun produces about that (140–150 dB) when fired, and an M4 is even louder at 150–160 dB. Note, too, that the decibel scale is logarithmic, so every 10-dB increase represents a doubling of the acoustic energy. That M4 is *twice* as loud as a 9 mm.

As weapons become more powerful, they produce more sound energy. The same goes for equipment—jet engines produce about 160 dB at takeoff. So, in warfare we've transitioned from muskets, which fire a subsonic projectile, to automatic rifles and heavy weapons that can deafen with a single shot.

In the firing range we took turns firing different weapons, standing close to each other due to limited space, and someone unloaded a full magazine from an unmuzzled M4 about a yard from my head. Although I had earplugs in, they didn't do much good.

Standard earplugs reduce sound by about 30 dB, but in this case the sound from the unmuzzled variant of the M4 was at least 170 dB, and my ear was exposed to 140 dB after the benefits of the earplug. Tests revealed a hearing band gap in that ear; certain frequencies were not audible.

Fortunately, due to some amazing medication, my hearing improved, but it didn't fully recover. I already had some damage in my other ear from scuba diving in my twenties, which took out some of my low frequencies. Between the two ears, I can hear the complete spectrum, but it took some adjusting. Still, it made me think: Is this the best we can do?

You have to be alive to have hearing loss.
—military aphorism

The number one disability for American veterans is tinnitus, a persistent ringing in the ears, from exposure to loud sounds including gunfire. As of 2014, nearly a million veterans were receiving disability for hearing loss and over a million for tinnitus. The United States spends more than a billion dollars a year on hearing disabilities, with most of those coming from the military.

The government could save significantly if those disabilities could be prevented in the first place, but the technology of hearing protection hasn't changed since World War II—you stuff something in your ear to baffle noise. Cotton and flexible foam have given way to modern plastics, but the principle is the same.

So we set out to design a better earplug.

There are two problems with standard earplugs. The first is that they cut the sound by only 30 dB or so, which, depending on circumstances, may not be sufficient. The second related problem is that they reduce all sound indiscriminately, including sounds you might want to hear.

Soldiers and law-enforcement personnel in the field are faced with the dilemma of whether or not to wear their earplugs on a mission. If they put them in early, the plugs can obscure the subtle sounds that indicate an enemy presence, a potentially fatal disadvantage. But if they don't have them already in place when a firefight breaks out, they may not have time to get them in place. Once bullets are flying, hearing loss becomes a low priority. It's rare that a solider has perfect information on when the shooting is going to start, and a single burst from a modern automatic weapon is sufficient to cause permanent hearing impairment, as I've experienced firsthand.

So we asked ourselves, what if there was an earplug you could wear all day that would allow you to hear normally but could alter itself to block damaging sounds instantaneously, all without requiring batteries or special training?

We took our inspiration from the shape-changing, liquid metal robot in *Terminator 2* and created an earplug from a new metamaterial with specifically designed nanostructures that are sensitive to environmental noise. Metamaterials derive their properties more from their custom-engineered, small-scale structure than from the underlying properties of their base materials. They can be engineered to have specific properties not found in nature.

These earplugs let normal volume sounds through unimpeded but activate in the presence of loud sounds. You can wear them all day knowing they won't interfere with your task at hand, but they will still protect your ears when needed.

This new material has a unique nanostructure that actually changes its shape as a function of sound. We teamed up with a scientist from CalTech who specializes in mechanical metamaterials to develop this material, building it in our lab, then testing and tweaking it at CalTech. And it worked beautifully.

The downside for our military customer was that the shape change was permanent, and they were rendered single-use, throwaway plugs. Once they'd been exposed to a loud noise, they didn't return to their previous state to let normal volume sound through. Our customer was okay with the earplugs being disposable but wanted to get several uses out of them first. In particular, they didn't want soldiers to have to replace them immediately after use.

So we blew up our old design and started again with a clean sheet of paper, which is not an easy thing to do. It's easy to get attached to an idea, especially a cool idea. We'd solved the earplug problem in a novel and technically interesting way, and it was tough to give that up. But attachment is the enemy of innovation, whether it's to a group consensus (groupthink), the way something is "supposed" to work (functional fixedness), an authority's expertise (expert thinking), or your own cool idea.

The new design also relies on nanostructure to baffle the sound but is more resilient, allowing it to be used more than once. The result was an earplug that removes 30 to 40 dB of noise in the presence of gunshots or explosions but otherwise has no effect on hearing. We call this new system Sound Guard.

//////////

Implications

Hearing loss is the third most common physical disease in the United States, behind arthritis and heart disease. Hearing loss can be difficult to detect as it happens, particularly in industrial environments where hearing is lost a little at a time. With the population aging in many countries, it's a big and growing business. In the United States, managing hearing loss with current technology costs $1,500 to $5,000 per patient. Multiply this by 48 million people in the country, and you can see the economic implications of this ubiquitous disability.

The science we brought to bear on this problem, which resulted in the Sound Guard Ear Plug, was produced using the concept of metamaterials—materials that can change as a function of the external environment. To manage noise, you need materials that sense the loud sound, then reduce noise exposure, then open back up when the noise abates. The implications of metamaterials are vast. Here are some examples:

- Buildings can be better insulated, more effectively ventilated, made quieter, or more open—the wall or the insulator can sense the external temperature or noise and adjust the wall to consistently produce just the right level of insulation or noise mitigation. This is particularly important as communities take advantage of transit-oriented development, and as city populations grow, more people will live in high-density, high-rise complexes. Managing noise between units and between floors will become a premium feature and more important to one's quality of life.
- The subterranean world we increasingly rely on could become more predictably manageable and harnessable, managing geophysical impacts on humans and society—including everything from earthquake prediction and compensation to groundwater extraction or high-performance fracking. Machines and systems could be further opti-

171

mized to integrate what's going on in deep underground processes.

- Clothing can adapt. For example, my wife and I hiked "the W" in Patagonia, covering up to 15 miles per day for several days. The weather changed from rain to snow, from warm to cold, from calm to windy in a matter of minutes. In the future, we could just wear one garment that could not only sense the external environment but also the condition of the person wearing the garment, so that when hiking uphill, rather than requiring taking off instead of changing layers, the single garment could automatically sense and adjust. This may seem whimsical, but it could have great applications in places like hospitals where regulating patient temperature is critical to patient recovery. The same would be great for garments suitable for space travel, not only regulating body heat but also integrated for physical sensing, adjusting for radiation shielding, and so forth.
- The Metamaterials could amplify and strengthen signal transmission, to deliver higher speed and quality of data and information. This itself has further implications for an untold number of industries.

/////////

Our overall mission at Oceanit is to address and solve the world's "impossible" problems. For your people to do amazing things, you must create an environment and expectation for greatness. Creating an environment where people can "bring awesome" to work is nontrivial. It requires a much more horizontal structure than the traditional command-and-control pyramid perfected during the Industrial Revolution, as taught in MBA programs and still employed by many businesses and the military.

However, a horizontal structure works only if it's hosted within the right culture. Creating a culture where talented people can be self-propelled requires creating an environment that supports open debate and dialog, particularly between different disciplines of study and fields. Experimentation is the great equalizer. Encouraging experimentation, learning from failure, and forging an approach into the unknown will yield amazing results. Whereas failure is viewed as an unacceptable outcome in most slow-moving or static business environments, it's viewed as an opportunity to learn in fast-moving environments, where rolling the learnings from failure becomes an asset that compounds in value with more experience. Moreover, the calculus of failure is that failing fast can often be more valuable than the time taken to reduce risk. Failing fast is another way to buy down risk, so when it's time for scale-up to occur, it's done well because there are fewer unforeseen problems.

Culture trumps strategy, but strategy can contribute to creating culture. At Oceanit we have established Strategic Principles—responsibility, diversity, *ohana*, embracing chaos—that endeavor to inform and empower people across the entire enterprise. These principles underscore what we have learned over and over: given the opportunity, most people want to do the right thing. Strategic Principles are really a statement about human nature. We believe most people, most of the time, want to do something good, if given the opportunity. This has also been our experience at Oceanit. It's a corollary to Steven Pinker's thesis in *The Better Angels of Our Nature* and his sophisticated explanation on why violence has declined. Just as Pinker does not dismiss the existence of violence, we don't dismiss the need for the right level of business supervision to prevent "bad actors" from causing trouble or to prevent self-inflicted business trauma.

Nevertheless, if you want highly educated people to bring their awesome capabilities to change the world, you need to give them the opportunity. Moreover, as I've said earlier, it is inherently more difficult to manage building a business to innovate, and it requires bright, flexible management with soft skills and flexibility as well as discipline. This is not how business managers are trained in college, where they teach how to "build a business to manage," versus innovate.

CHAPTER 10

How Emotion Drives Innovation

If you choose a job you love,
you'll never work a day in your life.
—attributed to Confucius

Richard Feynman, one of the great physicists of the twentieth century, whose name often goes hand in hand with Albert Einstein and Stephen Hawking, nearly gave up on physics altogether. Demoralized by the death of his wife and his work on the Manhattan Project, and with his best years as a scientist still ahead of him, he was ready to abandon one of the great passions of his life, saying, "I was convinced that from the war and everything else (the death of my wife) I had simply burned myself out."

Then one day he saw a student in the cafeteria tossing a plate in the air and noticed something odd. "As the plate went up in the air I saw it wobble, and I noticed the red medallion of Cornell going around. It was pretty obvious to me that the medallion went around faster than the wobbling."

That oddity sparked his curiosity, and because he had "nothing better to do" he thought about why that might be. When a colleague asked him what the possible significance of that could be, he said, "There's no importance whatsoever. I'm just doing it for the fun of it." Since he was too burned out for serious work, he decided to just enjoy himself, with no thought for the consequences. "It was easy to play with these things. . . There was no importance

175

to what I was doing, but ultimately there was. The diagrams and the whole business that I got the Nobel Prize for came from piddling around with that wobbling plate."

The "diagrams and the whole business" were the famous Feynman diagrams that students and researchers still use today to work out particle interactions at the subatomic level. He won the Nobel Prize in Physics (along with Julian Schwinger and Sin-Itiro Tomonaga) for his "fundamental work in quantum electrodynamics, with deep-ploughing consequences for the physics of elementary particles."

Play for Performance

Play lies at the core of creativity and innovation.
—Stuart Brown

What saved Feynman's career and ultimately won him the Nobel Prize was rediscovering the joy of physics—playing, having fun.

Play excites. It motivates. It encourages creative problem solving. But play is much more than an antidote to despair, it actually produces higher performance in knowledge workers of all types. Feynman was engaged in undirected, basic research, but the same principles are in effect across the board, including the applied sciences.

This, incidentally, is a problem our country is having pushing its STEM (science, technology, education and math) agenda in education. Schools believe that if they mandate STEM classes then their students will be better prepared for higher-paying STEM jobs on graduation. But they're missing a critical element—play. Kids are hardwired for play, but it gets systematically divorced from their educational experience with unsatisfying results. Not only do the students disengage from their classes, they lack the playful, experimental mindset to succeed in STEM careers.

In fact, I believe so strongly in the power of play and its necessity in disruptive innovation that the Oceanit mission statement, crafted more than thirty years ago, includes the phrase "where elements of work and play are indistinguishable." I wanted to create a work environment that would encourage creativity by allowing people to find some element of fun in their work. It should answer the question, If I wasn't concerned with money, what would I do with my time on Planet Earth?

Every day we ask people to do the impossible. What we need are people who think it's fun to try to walk through a wall where there is no door. No amount of money can motivate someone to do something they think can't be done. But a person who enjoys that challenge will be relentless in trying to make it happen.

Recently I encountered a book that captures much of my thinking on the benefits of play: *Play: How it Shapes the Brain, Opens the Imagination, and Invigorates the Soul*, by Stuart Brown. Of particular interest is how play activity or a playful approach to solving a problem evokes a more thorough exploration of the solution space and increases the chances of hitting on a viable solution that might otherwise escape attention.

As Brown says, "The genius of play is that, in playing, we create imaginative new cognitive combinations. And in creating those novel combinations, we find what works." He tells the story of a biologist who trained river otters to swim through a hoop to collect a treat. But the naturally playful otters weren't content to stick to what they knew worked. They tested variations, swimming through the hoop backwards or stopping halfway through and so on, to see if these would also qualify for treats. "By having fun and mixing it up, the otters were learning far more about the way the world works than if they had simply performed the initial task flawlessly."

We see this behavior all the time at Oceanit. Our most successful researchers are those who approach the work as play, and that's exactly the mindset that the Oceanit culture is designed to foster.

Brown concludes the story of the river otters with what is probably intended to be a joke, but it rings all too true: "The biologist ruefully noted that he had been trying for years to get his graduate students to use such playful investigation rather than rote learning and mechanical thinking in research." We find that too much time in academia stifles the creative impulse, discouraging creative risk taking in favor of playing by the rules and staying within the lines.

Transdisciplinary thinking, the core principle of Intellectual Anarchy, has a synergistic effect with play. People who come from outside a discipline are naturally more inclined to test the boundaries of that discipline and discover things overlooked by experts who have been conditioned to stay within the well-established boundaries.

There's an XKCD webcomic segment that neatly illustrates this idea of rules-testing play (true geeks know there's an XKCD comic for *everything*). Spoilers ahead for XKCD 1608: Hoverboard. If you haven't seen it yet, take a moment to check it out (http://xkcd.com/1608/).

Back? Okay. The comic presents as a simple, interactive browser game. A caption says "Here's a small game. Use the arrow keys to move." As you do, a stick figure on a hoverboard floats around, dodging walls and collecting coins.

It's possible, by intent or accident, to leave the bounded "play area," which has some artistic gaps in its border. If you do, an urgent message flashes in red, "Return to the play area."

If you continue, despite the warning, you discover a vast, sprawling play area, littered with in-jokes, secret chambers, and hidden passages. The *true* game dwarfs the initial "play area" by several orders of magnitude.

What's the message here? There are several possible takeaways:

- Those who play by "the rules" are missing out on most of what's out there
- Playing it safe inhibits discovery
- Obstacles we perceive as boundaries should be questioned and tested
- A big part of play is discovering what the rules really are

There are two other characteristics I want to single out as particularly important to disruptive innovation.

The first is courage.

When you're working on something truly disruptive, you tend to be so far outside the mainstream that the work will be ignored, disbelieved, or openly mocked. It takes courage to stick to your guns in that case.

Apple came under fire for removing the headphone jack from their latest iPhone. Senior VP Phil Schiller said, "The reason to move on: courage. The courage to move on and do something new that betters all of us." And for this, Schiller and Apple were widely mocked (e.g., "Never heard anything as ridiculous"—Mashable).

Apple has a pretty good track record of leading the charge in abandoning legacy features—serial and parallel ports, the floppy disk, the optical drive. Time will tell whether their decision to remove the headphone jack was a genuine and necessary innovation or a mistake, but Schiller was exactly right. Apple came under fire. They *knew* they would, and they did it anyway. That does require courage. It's not the same sort of courage as facing down a tank in Tianamen Square or charging a machine-gun nest, but courage nonetheless.

The second essential characteristic is emotion. We don't talk much about emotions in technical fields. There's a false perception that they're not relevant, possibly a distraction. In fact, they're vital. We've already seen how even a mind as great as Feynman's can be derailed by negative emotions. And positive emotions rising from a sense of fun and playfulness can lead to increased engagement, productivity, and creative thinking. Emotion is also critical to selling disruptive innovation. Any truly disruptive idea doesn't fit established molds, so it's easy to reject. Emotion is essential to convincing people to take a chance on something new. We see this all the time in the start-up world. People say "yes" with emotions and "no" with facts. At Oceanit, we harness emotion, backed with facts, to drive disruption, but it can also play out in the negative.

I recently met with a former Stanford professor who tried to advise Theranos founder and CEO Elizabeth Holmes to stay in school a little longer to further understand her idea of blood tests with a very small amount of blood. The then nineteen-year-old decided to launch Theranos anyway, raising more than $700 million and reaching a valuation of $10 billion at its peak. Investors caught up in the hype missed the warning signs that the technology wasn't ready and were wiped out when Holmes and Theranos were accused of fraud.

Likewise, Juicero raised about $120 million in capital on the promise it would do for smoothies what Keurig did for coffee, selling an expensive base unit and a recurring supply of consumables (packets of prejuiced fruits and vegetables). But Juicero's expensive Press device didn't actually work any better than hand squeezing the juice packets, and the company quickly shut down.

Changing Behavior through Information: Ibis Intelligent Networks

Behavior is what a man does, not what he
thinks, feels or believes.
—Emily Dickinson

Ibis Intelligent Networks is an example of the importance of an emotional connection to making Intellectual Anarchy work. Ibis began as a passion project by David, one of our engineers who had just joined us from California. He pitched the idea of a web of "smart sockets" for monitoring energy usage. You'd plug your appliances into individual smart sockets, which would in turn plug into your home electrical outlets. The sockets would create an ad-hoc network to communicate with each other and gather data to be displayed in an app. Each socket would be semi-intelligent and could be controlled independently but also share data with its nearest neighbors.

David had just joined us from California. He'd suffered through the energy crisis, watching oil prices peak and fall, and he wanted a device to track his own energy usage. And, too, he was excited about building intelligent, connected devices, a focus of his graduate work at Stanford in its Smart Product Design program (the program that immediately preceded their Design Thinking program). He wrote a white paper and made a presentation to our internal Innovation Fund within six months of arriving at Oceanit.

At the time we'd been doing research around healable, wireless mesh networks for the Department of Defense based on a secure, non–Wi-Fi, non-Bluetooth wireless protocol, so there was some synergy there. But as intriguing as was David's web of smart sockets, it didn't seem to have sufficient commercial application to warrant funding on its own merits. And David himself didn't have a background as an RF (radio-frequency) engineer—he just wanted an opportunity to play with some cool new technology. This actually made his project a perfect candidate for the Innovation Fund.

Our Innovation Fund serves a narrow niche. We fund only projects that look interesting and important but are too risky to be funded any other way. This is one way we encourage creative exploration and risk taking. We don't want our engineers and scientists to play too safe. We're asking them to find the edges of science and engineering. We choose ideas by applying a discipline informed by the Heilmeier Catechism, named after former DARPA director George H. Heilmeier (1975–1977), who developed the DARPA rubric for proposal evaluation. Successful DARPA proposals need to address a series of questions, summarized by the following:

1. What are you trying to do? Articulate your objectives using absolutely no jargon.
2. How is it done today, and what are the limits of current practice?
3. What is new in your approach, and why do you think it will be successful?
4. Who cares? If you are successful, what difference will it make?
5. What are the risks? How much? How long?

In the case of Oceanit's Innovation Fund, we review the same set of questions, but we also consider how, if successful, the new discovery could be used to bootstrap ideas of even bigger consequence. Simply said, if we can deliver a "proof point" to validate the kernel of a profound idea—one that would otherwise be considered unfundable or too crazy—we can use this proof point to raise additional funds to continue building the innovation. That's why Oceanit's Innovation Fund finances only things that are, by most standards, unfundable and too risky.

Funding the Unfundable

If at the first the idea is not absurd,
then there is no hope for it.
—Albert Einstein

Too many funds pursuing "innovation" play it safe, demanding a guaranteed win. But innovation, particularly disruptive innovation, is not something that comes with a guarantee. Recently I fielded a call from some investors in the Middle East looking to put together a fund. They had deep pockets and said they were interested in disruptive innovation, but at the same time they didn't want anything too controversial. They wanted the brand and the pizazz associated with disruption but also wanted predictable outcomes and certainty.

The Oceanit Innovation Fund (our internal innovation fund, now part of Oceanit Technology Ventures) is the opposite of this. If a project is a likely success, we'll fund it through normal development channels. We're looking for projects that are high risk and potentially high impact but have no likelihood of getting funded another way.

The Massachusetts Institute of Technology (MIT) runs a similar program. They're one of the few institutions out there funding the unfundable. The Amar G. Bose Research Grants, named for the late Professor Bose, provides $500,000, three-year grants to

MIT faculty for risky projects. Among the selection criteria are that the project is unlikely to be funded by traditional means and that research is intellectually adventurous. "Any truly groundbreaking research will likely be found to be risky, inappropriate, or unrealistic by many of the established practitioners in the field," Professor Bose's son, Vanu, says of the grants. "Historically, many of the innovative and groundbreaking advances in a field have come from people outside of, or on the periphery of, the particular field, since they are often able to bring a fresh perspective to the problems and ideas."

I should note that despite the risk inherent in these Innovation Fund projects, about 75 percent of them do eventually make it to market in some form or other.

The Technology Sherpa

We gave David a $25,000 budget and, more importantly, the time and freedom to purse the project. He designed and built the hardware—the first prototype was huge, the size of a laptop—wrote the control software, and demonstrated that it worked. And that would have been that—a cool project without a future, a learning experience, but not a product—until our "technology Sherpa," Ian, bought a Prius.

Ian has had a number of titles at Oceanit over the years, including marketing director and communication director. These days he considers himself primarily an advance talent scout, looking for the people we don't even know we need yet, but he's also responsible for finding ways to bring our innovative technology to market. We call him our technology Sherpa in honor of the Sherpa people of the Himalayas, legendary in their contribution to climbing the world's highest and most challenging mountains. It's a perilous journey to move research into the real world and, as with the Himalayas, if you attempt it without a Sherpa, you're in deep shit.

In 2010, Ian bought a Prius hybrid because he was interested in minimizing his carbon footprint. He noticed an efficiency meter on the dashboard that provided real-time feedback on how his driving habits affected his energy usage/fuel economy. Minimizing his energy consumption became a game for him. He accelerated and decelerated more gradually. He strove to maintain a steady 55 mph on the freeway. When he saw the impact that the efficiency meter had on his driving style, Ibis suddenly *clicked* for him.

Reports on electrical usage didn't thrill Ian, as they did David, who initiated the project. That information is already available, at a less granular level, from the electric company. An early prototype of Ibis featured a glowing colored cube as it's only output device. As long as energy usage was "acceptable," below a prearranged threshold, the cube glowed a reassuring green. But if energy usage crept over the threshold to "unacceptable," the cube glowed red. This unmistakable, clear signal—like the efficiency meter in his Prius—begged for immediate action. Ian realized that real-time feedback that Ibis provided could become a powerful force for changing people's behavior. And changing behavior is one of the most difficult challenges we face.

Educating the Market

The best way to predict the future is to create it.
—Alan Kay

Because we deal in disruptive innovation, we constantly wrestle with the problem of trying to move technology into a market that may not be quite ready for it yet. An evolutionary product is an easy upgrade—a computer with a bigger screen or a faster processor or a longer battery life. A revolutionary product requires more convincing; it requires educating the market. A computer with a brand-new input model, for example, is a tougher sell. The first Macintosh, with its WIMP interface (windows, icons, menus, pointer), graphical display, and mouse was initially perceived as a toy and took a long time to make inroads into the lucrative business market.

Once Ibis clicked for Ian, he knew we had a revolutionary product on our hands. The problem then became to convince others of that fact. We landed a small contract from the Office of Naval Research for our first pilot project, putting thirty sockets into military houses at the Marine Corps base in Kaneohe, on Oahu's Windward Side.

Marines are awesome customers, by the way. Unlike, say, the US Army, which is reliant on big forces and multiple specialties collaborating, the Marines have a "force of one" mentality. They're more independent and self-reliant. They improvise more than other services. And perhaps as a consequence, they're very direct about what works for them and what doesn't. They're very deliberate, very helpful. They'll tell you flat out, "this is crap" or "this is great."

We developed the software and created ten sockets through rapid prototyping. The tech worked, but people didn't respond the way we expected. We probably did fifteen different pilots for various organizations, trying to figure out the business case.

In one early project for the National Guard, we discovered that they had a number of old refrigerators wasting a lot of energy. Service members living in Hawaii don't want to drag their heavy appliances with them when they rotate back to the mainland, so they donate their old refrigerators. The National Guard had amassed nearly twenty donated refrigerators at this one site. They were old, poorly maintained, and far from energy efficient, but they were all running and drawing a lot of power. They had one from the 1960s chewing through thousands of dollars a year. By contrast, a modern refrigerator can use as little as $35 per year in electricity.

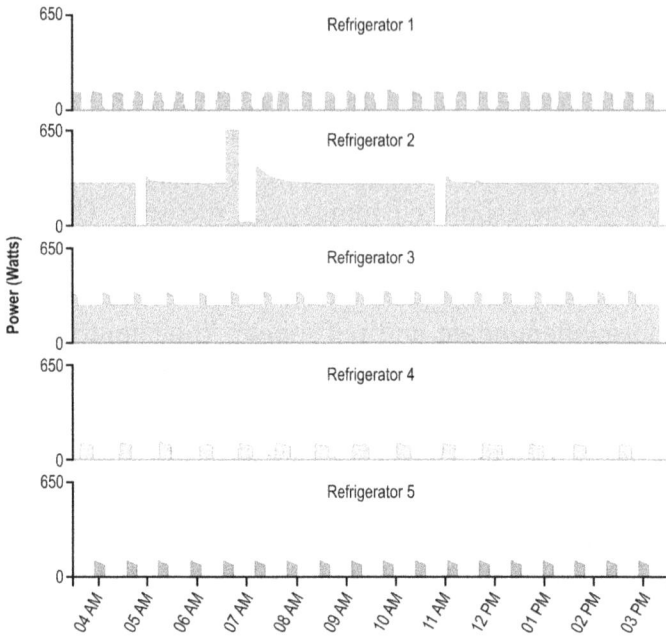

Fig. 10.1. Energy usage of several refrigerators, as part of a series of studies that included hotels and the University of Hawaii. You can see that the second and third refrigerator are having problems. A simple Artificial Intelligence system can simply identify these issues preemptively and schedule a maintenance call, saving cost and energy.

In figure 10.1, you see a comparative profile of the energy usage of five actual refrigerators at the University of Hawaii at Manoa, as monitored by Ibis InteliSockets. These refrigerators are all about the same size but vary wildly in their energy usage. A quick glance is enough to determine that the first, fourth, and fifth (charted in red, turquoise, and blue) are all operating normally. Each shows a modest energy draw for a few minutes about twice an hour as the compressor comes on. The compressor for the first refrigerator (red) activates twenty-eight times—twice as often as the third refrigerator (turquoise)—and costs $68 per year to operate versus $51 per year. The compressor for the fifth refrigerator (blue) activates eighteen times but for a shorter period of time and draws less energy when active, making it the most efficient at $35 per year. That's nearly half the cost of the first refrigerator, even though they have roughly the same profile.

The second refrigerator (charted in mustard) is clearly compromised. Its compressor is running almost constantly, as evidence by its nearly continuous, even power draw. This refrigerator costs $337 per year to run, more than ten times as much as the most efficient refrigerator, and it needs to be repaired.

The third refrigerator (charted in lime) seems to have a normally functioning compressor, as indicated by the small spikes every thirty minutes or so, but (I'm guessing here) it is simply old and inefficient, drawing a large amount of power even when the compressor isn't running. This refrigerator costs $197 per year to run and needs to be replaced.

Even this obvious win didn't gain us traction. It served as a one-time energy audit but didn't make the case for ongoing real-time monitoring. We explored the market. We needed to address bigger organizations with correspondingly bigger energy budgets that would see more value in tracking their usage: enterprises, hotels, universities, and schools.

We learned more with every project and confirmed that the value of Ibis InteliSockets was less about energy and more about information that impacts behavior. (The data is de-identified to look at overall usage patterns, not to track individual habits.)

In a way, it was similar to a project we did on the local sewer system. Sewers are a lot like any other network, with inputs, outputs, and the flow of, let's call it "data," through the system. For the sewage project, we put sensors on the pipes and monitored the flow of waste throughout the network. We then used that data to generate actionable information. We learned that usage was very regular; you could set a watch by it. Individuals are chaotic, but the community, in aggregate, is predictable. Outliers from the pattern of regular use became signals for leaks and illegal dumping. More significantly, we were able to determine which parts of the system were overutilized and which were underutilized and thereby create virtual capacity. The sewer system had been designed deterministically—that is, by looking at the number of users and building enough capacity to accommodate them, which resulted in enormous overcapacity in most segments. Not everyone with a toilet uses it at the same time. With our data we were able to build a statistical model of usage. For example, 50 percent of users flush their toilets between 9 a.m. and 10 a.m., another 25 percent between 10 a.m. and 11 a.m., and so on. The identified overcapacity becomes virtual capacity that can be repurposed, dialed into new infrastructure that costs little but has a huge return on investment, or used to delay construction of additional future capacity.

Our work with Ibis was similar. A large part of the value of Ibis is the intelligent analysis of data captured by the network of smart sockets and the ability to shape electrical usage by turning on or off individual sockets based on that data. Most office buildings, for example, are empty nearly two-thirds of the day, yet the office equipment continues to draw power around the clock. Even devices with a power-saving mode typically still draw a trickle charge. Ibis allows you to intelligently monitor and control these devices, detecting usage patterns and completely shutting down power-hungry copiers, printers, monitors, and so on when they're not needed. A typical office can save 15–20 percent on energy costs.

Ibis also allows you to avoid expensive demand charges by managing your devices to keep usage below a designated cap. Power companies are limited by their peak capacity, so they charge different rates for energy usage during peak and nonpeak hours and "demand rates" for businesses—meaning their rate for the billing cycle depends on the maximum electricity businesses use in any given fifteen-minute period. A company that uses an average amount but incurs a one-time spike can get hit with massive demand charges (like going over on your data limit with your phone). Ibis allows you to avoid this scenario by automatically powering down noncritical systems as usage nears the threshold.

For businesses, avoiding peaking means saving money. For power companies, it means avoiding brownouts or blackouts or deploying massive capital for new capacity to address peaking. They're not trying to punish or gouge their customers with demand charges but to shape the demand to what they can accommodate. Peak demand is the most important constraint they have to manage. This gets even more complicated with an increase in renewable energy, which mostly comes from inconstant sources like wind and solar. Now you have to plan around cloudy weather and winds from different directions in addition to your other constraints.

All this leads to the obvious question: If we can create virtual capacity in a sewer system and in an office building's electrical usage, can we do the same for an entire power grid? Maybe; we're actually talking to HECO (the Hawaiian Electric Company) now about a pilot project to do exactly that. Consider the possibility of a HECO–distributed InteliSocket that earns the user a 10 percent discount on all electricity used by the attached device, with the caveat that it may be shut down during peak energy usage, or that it can be intelligently managed to provide a staggered start to mitigate peaking. For example, schools around Oahu begin their day at the same time and turn on their air conditioners at the same time, causing a large, predictable energy draw. The schools care only that the AC is on early enough to cool classrooms and offices by the time students and staff arrive. If HECO had the schools' approval to automatically turn on their AC up to 20 minutes early, then they could turn on 5 percent of the air conditioners every minute over that interval, rather than all at once, smoothing out the energy draw from initial startup.

At this point, the question for us was not whether to proceed with Ibis but how best to proceed—internal funding and development? Licensing? Corporate codevelopment? We knew we didn't want to get into the consumer space (the "gadget market") and compete with Apple, Belkin, and others. Ultimately, we chose to spin out Ibis into its own company. (See chapter 12, "Mind-to-Market," for a closer look at these various options and their tradeoffs.)

Conceiving a product like Ibis and then driving it to market takes time, energy, and effort. The stronger the emotional connection to the project, the easier that effort is to sustain. We see this over and over at Oceanit, and it's one reason we weight project selection toward ideas that we feel will have a significant, meaningful impact on the world if we can pull them off.

The Long Pipeline

We originally developed the mesh networking technology used in Ibis for a completely different (military) application. The fact that we reused it in a commercial, energy-saving context is a useful illustration of three Intellectual Anarchy principles: the journey of discovery, the avoidance of functional fixedness, and the long pipeline:

- *Journey of Discovery:* When you ask an interesting question, you'll get an interesting answer, although it may be hard to predict the outcome in advance.
- *Functional Fixedness:* The cognitive bias that prevents you from seeing the utility of an object for something other than its intended purpose.
- *The Long Pipeline:* Some discoveries may not pay off immediately; it can take years. A short-term outlook prevents you from pursuing them, but a long-term outlook lets you fill your pipeline with innovation that pays massive dividends down the road.

Taken together, these three principles suggest a strategy. We ask interesting questions not knowing where they will lead us. Sometimes the result will be interesting, but the market won't be ready. These ideas can be revived later or adapted to completely different problem domains than originally intended.

This is exactly what happened with Ibis. The mesh network

190

technology was originally developed to help soldiers communicate in urban environments. Radio is basically line of sight, so when soldiers are conducting a house-to-house or room-to-room search in a village or cavelike structure, it's easy to lose contact with the rest of their team, and that can be fatal. Our solution was to develop small, cheap radios that could be deployed automatically to maintain a link of effective line of sight back to the team. The radios were smaller than a quarter and cheap enough to be disposable. Soldiers would wear a pouch of these little radios and they'd drop on their own, like bread crumbs through a hole in a bag, whenever the signal got weak (we coated them in silicone rubber to survive the fall from the dispenser). Despite their size, these radios had enough intelligence built into them to form a self-healing mesh network that could maintain communications even if a single node were lost or displaced.

The military liked these, but their adoption cycle is very, very long: years. In the meantime, we went looking for other markets because we thought they were a cool solution. We tried first responders—police and firefighters in particular face similar challenges on the job—but they didn't have the budget. We shelved the technology, only to revive it when we were developing Ibis.

As another aspect of the same project, we developed wireless sensors to detect enemy combatants behind concrete walls. We measured heart rates, uniquely identifying individuals and comparing those against known (friendly) signatures, leaving unknowns as potential unfriendlies. This information could be distributed to the team over the same network. We recycled this technology into Ibis as well as part of our intelligent usage analysis. The problem domains couldn't be farther apart, but the physics and math turned out to be the same. Finding information and hiding information are opposite sides of the same coin.

This sort of cross-pollination is a huge part of what we do, and it's one of the reasons we're so eager to throw ourselves into anything that looks worthwhile even when the short-term payoff isn't immediately evident. Because you never know when some technology you've developed for one application turns out to be the missing puzzle piece of your current project or the springboard for an entirely new disruptive innovation.

//////////

Implications

It has been estimated that there were about 5.6 million commercial buildings in the United States in 2012, comprising 87 billion square feet of floor space. By providing fine-grain monitoring and control over energy usage at the plug level, Ibis can achieve the following results:

- *Reduce the cost of energy.* This is a big win for businesses, which can cut their energy spending by up to 30 percent. Ibis is smart enough to determine when it can shut down a light or piece of equipment without disrupting users. Today, just saving 10 percent of commercial building energy use would reduce carbon by about 20 million tons per year, the equivalent of pulling 15 million cars off the road.
- *Reduce the usage of energy.* In addition to cost, governments are interested in reducing energy usage to meet green goals and delay building new infrastructure.
- *Monitor device health.* With plug-level usage info, Ibis's intelligent monitoring software can identify faulty and inefficient devices that need to be repaired or replaced. It's possible to determine, just from energy use patterns, whether an appliance is working the way it's supposed to—that is, is it healthy, or is it likely to fail soon? A refrigerator, for example, runs its compressor on a regular on/off cycle, drawing a varying but predictable load. If the compressor stays on, that's a problem—maybe it's the compressor itself, maybe you have a coolant leak, maybe the relay is busted. But the increased load from the com-

pressor running nonstop could be detected at the socket. Smarter software, with AI bolted on, could not only detect the increase but flag it as something more significant than merely increased usage, prompting the user to take action. Without this sort of intervention, the faulty fridge might burn hundreds of extra kilowatts until the problem is discovered by other means.

- *Reduce the risk of electrical fires.* Faulty equipment is also the culprit in many electrical fires. Identifying and taking these devices off-line makes buildings safer and reduces insurance premiums.
- *Better utilize expensive assets.* In places like medical centers and universities where expensive equipment is required to execute research, some equipment can be shared with other elements of the enterprise. However, it's difficult to know if equipment is actually being used or is idle. Ibis, augmented with AI, will know the difference.

//////////

The emotional content of what we do is a critical element of how we work. We see it all the time as people are excited to come to work and take a crack at a big issue or problem. They are driven by the excitement, opportunity, and engrossing stimulation from pursuing a mystery or challenge. Just like beauty, you know it when you see it, though we struggle to adequately describe how this works. Nevertheless, as we look around to consider how others describe the connection between cognitive and emotional abilities, there are some interesting findings and observations.

Just as Angela Duckworth has found with her work on "grit," we have learned from projects like Ibis that the power of passion and perseverance, combined with the willingness to execute and deliver, produces amazing results. When we see otherwise very talented people fail to reach their potential, very often we find they lack this emotional element. They lack "grit."

We look for this quality when we recruit by asking people how they play—what hobbies or activities they pursue that are just for fun. This provides insight into the emotional range people are comfortable with. Although not always definitive, it gives us an indication of which people can consistently deliver disruptive innovation—those who can endure the physical and emotional challenges of pushing beyond their comfort zone.

Psychologist Mihály Csíkszentmihályi, who pioneered the concept of "flow," defined it as the state of losing oneself in a task or activity in which awareness and action become one and the same. In Csíkszentmihályi's chart of eight mental states, flow lies between arousal and control and opposite apathy. Apathy lies between worry and boredom. Worry may hold the seed of motivation to effect change, but it's a pretty weak diet. Grit can safeguard us from these unproductive mental states when we are tackling large problems that are ill defined and can seem insurmountable. Whether you are looking for a solution to a large problem or just looking for the most innovative way to do what you are already doing better, grit will help you get there, and staying in the flow will keep you going.

The challenges of climate change and the prescriptions for drastic lifestyle changes to thwart its biggest risks sometimes seem designed to engender worry and apathy. This is where a technology like our Ibis Intelligent Networks comes in. The mash-up of artificial intelligence and energy monitoring promises an enormous opportunity to reduce energy without the "inconveniences" of a green agenda. Many of us early adopters bought into a green agenda and do things as simple as turning the lights off when leaving a room or working in LEED (Leadership in Energy and Environmental Design) buildings and offices, but that's just the beginning of what's possible. Changing behavior is difficult. Individuals have their habits and priorities, and most of the business/commercial world in which they work has not made this change either. Their key mission is business driven; they are focused on survival, competition, and profit first, and only then do they consider the environment.

However, the convergence of AI and fine-grain monitoring and control over energy usage at the plug level will transform this. We will become more energy efficient, reduce greenhouse gasses, reduce the risk of climate impact, and save money, too. Today Ibis has simple, secure, wireless commercial connectivity and monitoring. The next step is to integrate AI into the system so that it won't inconvenience people who are focused on doing their job. This is where most of the world lives—and paradoxically where the greatest impact will be made.

PART IV
Deliver

The last mile. Innovation that does not see the light of day is like one hand clapping, an interesting rhetorical exercise, but unsatisfying to most. Delivery requires shifting cultural gears.

Two hands clap and there is a sound.
What is the sound of one hand?
—Hakuin Ekaku

CHAPTER 11

Deep Science to Human-Centered Design: Finding a Common Language

Innovation in the sciences is always linked in some way,
either directly or indirectly, to a human experience.
—John Maeda

At the beginning of this book I define Intellectual Anarchy as "a counterintuitive strategy for persistently producing disruptive innovation through the interaction of deep science and human-centered design."

Along the way I've explained how we use transdisciplinary thinking to avoid the expertise trap, how we defeat cognitive biases like groupthink and functional fixedness by building diverse, interdisciplinary teams, and how we gain an innovation advantage by connecting emotionally with our work, fostering a sense of play, and getting hands-on while making actual stuff. Now it's time to dig into the final terms of our definition of Intellectual Anarchy, "deep science" and "human-centered design," and examine how they can be harmonized for effective innovation.

Let's start with just *science*, which is frequently misunderstood and often confused as a synonym for technology. Science is the discipline through which we learn about the observable universe, proceeding via investigation, exploration, and experimentation. Science gives us a method for understanding the world around us.

Suppose, for instance, we're interested in improving the quality of the coffee we brew every morning. We can come up with a bunch of variables that we think are related. Grind, water quality, and brew time are obvious candidates, but maybe we also suspect that day of week and time of day are important. Then we run experiments, collecting and analyzing data. And we find that the finer the grind, the better the taste, up to a point where the trend reverses; that distilled water tastes better than tap water, but time of day has no effect whatsoever, and so on.

This is science. We learned something about our world through empirical evidence. At the same time, it's fairly shallow science. We improved the taste of our morning coffee, but we haven't discovered anything earthshaking. Anyone with high school level math, a coffeemaker, and a bit of patience could achieve the same result.

Basic science is routine. It's not deep, but it's useful. That's why everyone should take science in high school. It enables application innovation, consisting largely of tweaking known principles, such as using hotter water to produce better-tasting coffee.

Deep science, by contrast, wrestles with first principles and fundamental concepts. Consider the vast fields of application opened up by the discovery of X-rays or the invention of lasers. Deep science is inherently disruptive and results in disruptive innovation.

This is why we have an affinity for new PhDs when hiring at Oceanit. A doctoral student has a proven track record of not only mastering a field but adding something new by bringing first principles in science to an unsolved problem. For a lot of people, this is difficult. You have to go right to the edge of a field and then push beyond it. And it's not always easy to understand where the edges might be. You have to examine the work of everyone who's gone before you, read all the papers, and become an expert in the field—and then find a way to add something new.

The challenge then becomes getting this innovation out of the lab and into the world. Unfortunately, we've found that the first market inclination of scientists and researchers is almost always wrong. As specialists, they've developed a detailed understanding of how a particular technology works, which challenges their understanding of its broader applicability in the market. They can't see the forest for the trees. This is a lesson we learned early on doing ocean surveys with the MOP program.

Eel Goggles in the MOP

The Marine Option Program (MOP) is an interdisciplinary certificate program for undergraduates offered at the University of Hawaii for any students interested in the ocean. We worked with MOP students to perform benthic habitat surveys—surveys of marine flora and fauna on the ocean floor.

To conduct a survey, a team of two scuba divers marks out an underwater grid of one-meter squares and counts the number of coral, fish, invertebrates, and so on, in each square.

We found that if we brought in specialists, the results of the survey would be skewed toward their area of expertise. An eel expert, for example, would show that there were a lot of eels in the region. We couldn't make realistic comparisons from location to location, habitat to habitat, because all the observations were weighted toward eels.

MOP students, by contrast, produced unbiased results that could be used to compare marine benthic habitats in many locations, year after year. To qualify for the program, the students had to demonstrate a high level of skill, and then they all received the same training, which covered the entire marine ecosystem. This reliable standard of comparison was critical to understanding the impact of human development on the marine ecosystem.

Likewise, when taking a technology to market, we favor technological generalists who have the observational skills to acquire an unbiased understanding of what the market needs—what it is actually saying versus what we would like it to say. For disruptive technology, this requires more than a market survey. Observation, empathy, and active listening are critical to finding the right first markets.

We don't see this as a failing of the individual scientist, merely a trapping of human nature. But we have to work around it if we want to bring disruptive technology to the market. That's where human-centered design comes in.

Marrying Science to Human-Centered Design

Innovation doesn't just come from
equations or new kinds of chemicals,
it comes from a human place.
—John Maeda

Design powerhouse Ideo defines human-centered design as "a process that starts with the people you're designing for and ends with solutions that are tailor-made to suit their needs." Human-centered design is "all about building a deep empathy with the people you're designing for; generating tons of ideas; building a bunch of prototypes; sharing what you've made with the people you're designing for; and eventually putting your innovative new solution out in the world."

What good is innovation if it languishes in the lab? Or if the people for whom it is intended find it cumbersome to use or simply not worth the effort? The goal of human-centered design is to concern ourselves with the needs of the end user much earlier in the process. We have to start with the right question, not jump to a convenient answer because of our predisposed preference. What does the candidate customer need? In our case, the customer may be anyone from a soccer mom to the Department of Defense, but the principle is the same.

Marrying these two concepts, deep science *and* human-centered design, is essential for transitioning to the market from the lab, and it is one of the defining characteristics of Oceanit as a "Mind-to-Market" organization, distinguishing us from other, seemingly similar institutions that focus primarily on one or the other, although there are good reasons for this, as we'll see. (I explore the concept of "Mind-to-Market" in detail in the next chapter.) On the one hand, you have universities and national laboratories that focus on the science side of the equation, expanding the boundaries of knowledge but rarely transitioning out of the lab to have an impact on the lives of everyday people.

For example, the recent discovery of gravity waves by the Laser Interferometer Gravitational-Wave Observatory (LIGO, MIT, and Caltech) was a milestone in physics and astronomy—sufficient to merit the 2017 Nobel Prize in Physics—but the practical payoff, if any, is probably decades away. On the other hand, you have design firms like Ideo that focus on the human side of the equation, with practical innovations based on existing technology to improve people's lives in small ways. A modular shopping cart may streamline or enhance your grocery-store experience but isn't going to change the world.

There are pros and cons to tackling deep science and human-centered design in the same organization. The advantages include opening up the possibility space enormously. This broader perspective allows us to think about what we *should* do, versus what we *can* do. Our goal is to tackle the world's most important problems, not necessarily the easiest or most short term. It also allows us to deliver revolutionary products that have a huge impact. The disadvantages are that it's really, really hard. We have labeled these two zones as "Mind" or "Market." Both of these areas are challenging, and it's difficult to master one, let alone both. Individuals who excel in science or design specialize in them to such an extent that they have a hard time even communicating with their counterparts across the aisle.

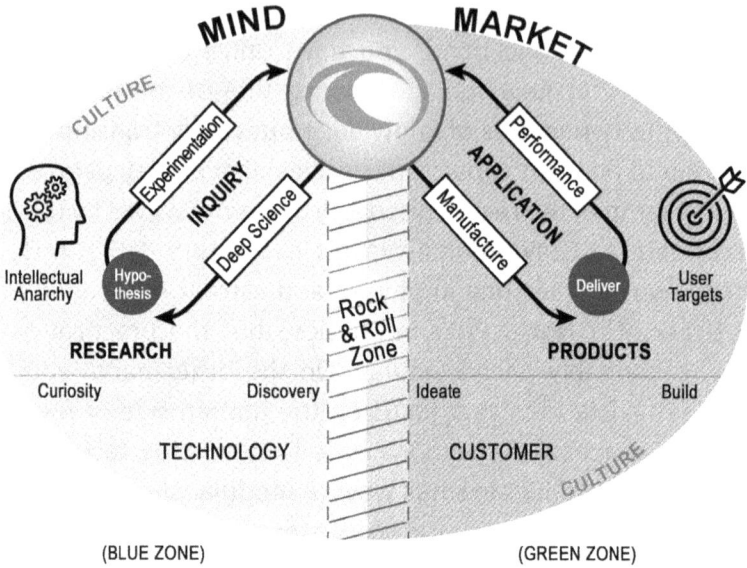

Fig. 11.1. The Mind-to-Market business model is challenging because of the cultural differences between the "Blue Zone" of deep science and the "Green Zone" where technology is delivered. Those two zones are connected by a naturally chaotic "Rock & Roll Zone"—innovation is messy. Typically, the original innovation may have little to do with where it actually ends up, since the market may have different ideas on how the technology is best used by its early adopters.

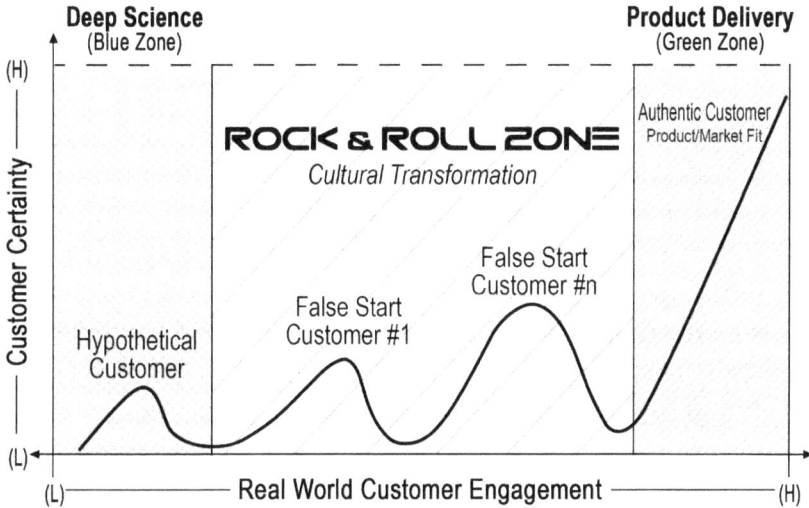

Fig. 11.2. The Rock & Roll Zone bridges the gap between the "Mind" Blue Zone (science and discovery) and the "Market" Green Zone (application, product delivery, and customer satisfaction). It's a highly nonlinear and chaotic environment that requires an inquisitive and agile mindset—sometimes compared to the challenge of looking for a light switch in a dark room. People who live in this zone are either having a blast or terrified. Rock & Roll Zone dwellers must be excellent listeners as well as great communicators.

At Oceanit, we refer to this culture clash as the "Mind" Blue Zone versus the "Market" Green Zone (see figures 11.1 and 11.2). The Blue Zone, on the left, is the culture of deep science. The Green Zone, on the right, is the culture of the market—humans and society. Blue culture values inquiry and discovery and employs the tools of science—mathematics, experimentation, and the scientific method—to deliver technology. Green culture values application and user/customer satisfaction and attempts to take newly discovered insights and technology to the market with its own set of tools: empathy, user feedback, affordability, supply-chain management, and so on.

The interface between these two cultures, the cross-hatched area, is what we call the Rock-and-Roll Zone. This is where worlds collide and cultures clash, where people from both sides are thrust out of their comfort zones and forced to deal with unfamiliar concepts and terminology. This is where things get messy—but it's also where they get interesting.

To those in the Blue Zone, the task of their counterparts in the Green Zone seems easy. Researchers in the Blue Zone have already done the hard work of proving the improbable, creating a seemingly impossible technology; all that remains is to bring it to market. But to those who live in the Green Zone, understanding the details of how that technology will impact humans and society is equally challenging. Factors like usability, which seem obvious from the perspective of the Blue Zone, can bring immense challenges. Indeed, one of the hallmarks of usability is that using it properly *should* seem obvious. But arriving at that "obvious" solution can be a lot of work. The challenges of the Green Zone can be every bit as difficult as those of the Blue Zone, particularly since the Green Zone deals with human beings and the diverse subcultures, behavioral economics, and ethnography of various industries, from the roughnecks and roustabouts of the oil industry to the nurse practitioners and doctors of the health care industry, and more. The well-known misquotation attributed to Ralph Waldo Emerson—"Build a better mousetrap, and the world will beat a path to your door"—might have worked in the 1800s; today it's more complex.

The ability to transition between these two cultures, Blue and Green, science and design, requires exposing each to the other. Delivering technology requires an enormous effort, not only in making that technology work in the first place but understanding the appropriate business model, usability, and quality systems that are required to make something at scale. To facilitate communication between the two sides, we rely on a common language: Design Thinking. We've learned to use the language of Design Thinking to create a common framework for understanding between the two cultures.

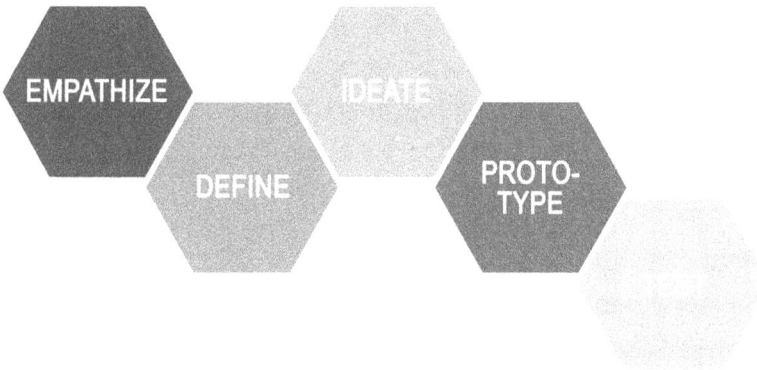

Fig. 11.3. The Design Thinking process was codified and taught at the Stanford d.school by David Kelley and his team. We started collaborating with Stanford over a decade ago to launch "Design Thinking Bootcamp Hawaii," which has infused creativity and innovation tools into Hawaii's schools, as well as the University of Hawaii. At Oceanit we use Design Thinking as a common language that connects "deep science" to "human-centered" technology delivery, enabling us to connect two very different innovation subcultures.

A quick glance at the Design Thinking process diagram reveals a process that's mostly familiar to engineers, with one notable exception. The final four stages of the process—define, ideate, prototype, test—are standard procedure for any engineering development project. Define your problem, come up with ideas for solutions, prototype those solutions, and finally, test the result.

However, the first stage, Empathize, is new. "Empathy" is a word that seems more at home in a yoga studio than in a science or engineering lab. You don't need empathy to bend sheet metal or mix a solution. But you do need it to create usable solutions for actual humans.

Engineers are notoriously bad at designing products and interfaces for nonengineers. They assume their end users will be people like them—people who enjoy reading manuals, who have sophisticated domain knowledge, who aren't afraid to experiment, who are willing to drill down through five levels of menus, and so on.

As a quick example, if you've ever set up a Wi-Fi router, you were probably presented with a dialog box that asked you to choose an encryption protocol with a choice of WEP, WPA, or WPA2. The engineer who designed the system, if they were in the room with you, could tell you that WEP has been deprecated, WPA has been superseded by WPA2, and that, unless you have a specific and compelling reason not to, then you should be using WPA2. Yet the list is presented in the opposite order, with WEP first. The worst possible option is the default if you just hit Enter, and there's no guidance offered at all.

Why? Because to engineers this makes perfect sense. They value power and configurability over simplicity. They've preserved the opportunity to select an obsolete protocol for the rare instance where it makes sense. Oh, and the various options are presented in lexical sort order—obviously.

But the average user just wants to get on the Internet, as quickly and securely as possible. Choosing an encryption protocol is an annoying extra step that provides no value. Worse, it diminishes value because it offers the option to choose a less secure protocol for no benefit.

This is a failure of empathy, a failure to deliver actual value to the user. Smart people of a certain stripe tend to be bad at listening. They already know more than most people, especially within their area of expertise, and they're less interested in things outside their area of expertise. But listening, and empathizing, are critical to delivering genuinely usable technology.

As always, the challenge is magnified when that technology is disruptive. To succeed in the market, disruptive tech has to nail the sweet spot at the intersection of innovation, desirability, and viability. This is the difference between creating something new but isn't affordable or scalable or between something that works but isn't adopted by consumers. This is the difference between delivering the Apple iPod that succeeded on all three and the Elger Labs MpMan F10, which managed one of three at best. The MpMan was the very first Mp3 player to hit the market, nearly three years before the iPod—but it was a graceless brick the size of a tape-based Walkman, with just 32 Mb of memory, enough for maybe twenty songs. It disappeared without a ripple. The iPod, by contrast, sold nearly 400 million units and became the best-selling music player in history, until it was disrupted by the iPhone.

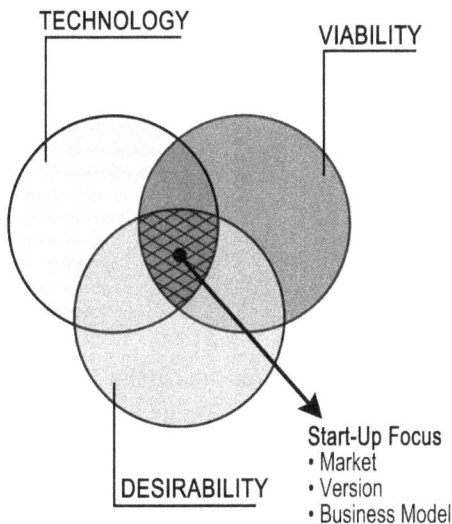

Fig. 11.4. To spin out a technology, three key issues need to be carefully considered, including technology, viability, and desirability. Opportunity lies at the intersection of technology, viability, and desirability that yield the market, version, and business model.

The common language of Design Thinking facilitates the kind of cooperation and collaboration necessary to hitting the intersection of technology, viability, and desirability. Moreover, it enables a mutual respect between the Blue and Green cultures. Both need to appreciate the challenges involved in bringing new scientific discoveries to market in ways that positively impact humans and society.

Contactless Medical Sensors: Two Sides of Human-Centered Design

Socrates said, "Know thyself." I say, "Know thy users."
And guess what? They don't think like you do.
—Joshua Brewer

In a room filled with high-tech gear—hard plastic cases stuffed with electronics and bristling with sensors and controls—sits an incongruous teddy bear. His name is Teddy, and he is the immediate point of focus for most visitors to Oceanit.

The room is our Technology Petting Zoo, a place for visitors to get close up and hands-on with some of our technology. They can see it, touch it, try it. Our nanotech surfboard lives here, as do demos and prototypes for the Q-Dive helmet, Ibis, and HeatX.

Technology Petting Zoo

Visitors to Oceanit, unless they're in a technical field, can have a difficult time understanding our products and what we do. It's too far removed from their everyday experience. So we put our products into a Technology Petting Zoo, which gives them a hands-on opportunity to interact with our products as way of understanding and connecting with the technology. But their interaction changes our perspective, too, as we see how they respond to it.

We have a demo of our VIPA video analytics software platform that people love to play with. We can take a video of a San Francisco freeway and subtract the cars, leaving just the road, or eliminate the vibration in a video shot from a jet taking off from an aircraft carrier. People are at their most creative when they play (see chapter 10), constantly testing boundaries and limits.

Visitors will often ask great questions inspired by and about how to use the tech, sometimes leading to new insights. We love getting a fresh perspective on a product, and no one has a fresher perspective than someone encountering it for the first time. We had a visitor ask about our SOAP electroplating process, wondering if we could shove it through a pipe to plate the inside—something we hadn't previously considered. And a question about Anhydra led to us putting it on heat exchangers on military aircraft.

Most of the equipment in the Petting Zoo radiates an aura of cool utility and efficiency, but Teddy invites a different response. He's friendly, approachable, huggable. And that's the point.

Teddy is actually filled with sophisticated, contactless medical sensors. He can detect the vital signs—heartbeat and respiration—of the child who picks him up and transmit them wirelessly to a display. Neither the display, nor the nurse that reads it, has to be in the same room. Suddenly a scary doctor visit is transformed into playtime. This is what we mean when we talk about human-centered design: technology that has compassion for those affected by it.

The LifeBed: Preventing Medical Distress

Children aren't the only ones that are put off by visits to doctors and hospitals, though they're more open about their feelings. Doctors, dentists, and needles all feature in lists of common fears. Our collective distaste for medical treatment comes at a very real cost. The price of noncompliance—failure to seek medical intervention in a timely fashion or to follow a doctor's recommendations—is estimated at somewhere between $100 and $300 billion *annually*. Anything we can do to ameliorate that fear and distaste is a huge win.

At Oceanit we developed contactless medical sensing technology, which has since been spun off into an independent, venture-backed firm called Hoana Medical. Teddy was an experiment using the same technology, but Hoana's major product is the LifeBed.

The LifeBed resembles something out of a *Star Trek* sick bay. The patient lies on an uncluttered, linen-covered mattress. No sensors are in evidence. No wires, clips, cuffs, or electrodes are attached to the patient. And yet a monitor beside the bed displays their vital signs.

The sensors are embedded in a thin coverlet that can be fitted to any standard hospital bed. They work on the piezoelectric principle, generating an electric signal when they are flexed or compressed, and they are sensitive enough to pick up the rise and fall of the patient's chest as they breathe and the beating of their heart.

This contactless technology delivers some major improvements over conventional technology, including the following:

- Increased patient compliance—nothing is required other than that they remain in bed
- More reliable signaling—no electrodes to disconnect
- Improved patient comfort—no restrictive wires
- Reduced setup time—nothing needs to be set up, attached, or calibrated, and the bed activates as soon as the patient lies down

The LifeBed delivers what we term Intelligent Medical Vigilance. It's not a diagnostic tool so much as a way of detecting a patient in distress and alerting a nurse. A study done in the United States estimated that 200,000 people die every year from preventable medical errors, with the bulk of those attributed to a shortage of qualified nursing staff. Typically, nurses have time to check on a patient only every two to four hours. If a patient in a general ward (not the ICU) suffers respiratory or cardiac distress between visits, they may die before anyone comes around to check.

The Science behind Hoana Technology

Anyone who surfs develops an appreciation for and intuition about ocean waves. Contrary to what most people think, waves don't move water across the surface of the ocean. Instead, the surface shifts up and down as the wave passes through it. When a surfer catches a wave, they're not being pushed by a wall of water but rather sliding down a hill that's risen up where they are (until the wave breaks; as waves get closer to shore, they're affected by friction with the ocean floor: breaking waves *do* move water).

Waves transmit energy and information through the ocean. They may look chaotic, but the underlying pattern can be teased out with signal-processing techniques, as I learned analyzing ocean wave data for work on coastal structures.

The human body is about 60 percent water. As the heart beats, it also transmits waves through the body that we can read using piezoelectric polymer sensors, which generate electric signals in response to mechanical stress such a flexion or compression. We can then analyze those signals using the same signal-processing techniques. The idea is similar to ballistocardiography, an early attempt to understand congestive heart failure that dates back to the nineteenth century. The patients were isolated from their environment, supported on an air bearing, and linked to an ink pen that would record squiggles on a paper drum from the vibration of their heartbeat.

The impetus for the original device was my son, who was born with a heart arrhythmia, putting him at an elevated risk for SIDS (sudden infant death syndrome). As a first-time parent, I was extra-sensitive. The hospital sent us home with a monitoring device—a chaotic bundle of wires and electrodes that was difficult to set up and didn't work very well. Attaching the electrodes while he was sleeping would wake him up. And as he got older, he'd move around in his sleep and wake up with wires wrapped around him. My son, fortunately, turned out fine, but a friend's child died from SIDS around that time. I thought—that's preventable; there has to be a way to track an infant's breathing and heartbeat without hooking them up to anything. The death of my friend's baby, that pain, was what started the journey. We put together a working system within a few months.

We went to the NIH (National Institutes of Health) for funding. They were looking at SIDS but didn't have much money for funding. As is often the case, there was money for soldiers but not for babies.

Our next major development effort came through a US Army contract looking at ways to improve triage. The first hour after an injury is known as "the golden hour." If you can get a soldier the care they need within that hour, you vastly increase their chance of survival. Medical teams go out in Blackhawks to deal with casualties and the wounded. They needed a way to monitor injured soldiers during helicopter transport. If you've ever been on a helicopter, particularly a military helicopter, you know that the noise and vibration are intense. You can't track anything in flight. Screaming "Are you okay?" is the best you can do. There's so much vibration that electrodes detach. You have wires everywhere. But our system doesn't rely on electrodes, so it's not subject to disconnect. It also has the advantage of working through battlefield uniforms and body armor. Sometimes you can't take those off. Our system works through all of those.

We were worried about vibration, though. We developed and tested the system with a paint-can shaker from a hardware store, wired up to a variable power supply. We put a volunteer on an army cot, put the cot on the paint shaker, and shook it at various rates to ensure our systems could always determine a patient's basic vital signs, even during extreme shaking. Lying on a paint shaker is an intense experience! That first electronics package was the size of a refrigerator, but it worked. The next iteration was small enough to fit on a Blackhawk, and we conducted a demo with volunteer soldiers wearing different types of gear: biochem warfare (MOPP) suits, flak jackets, and so on.

Because the sensors don't rely on a wired connection to pick up the patient's vital signs, they're not subject to disconnection—the main advantage they have over an EKG in a helicopter transport scenario. But by their very nature, piezoelectric sensors are highly susceptible to vibration. Every bend and flex generates a signal, which would seem to make them a worse choice for a helicopter. We got around that with some clever signal processing. We dedicated one sensor to the helicopter and one to the patient. By subtracting the helicopter signal from the overall signal, we were left with a pure patient signal.

Taking the LifeBed to Market

The success of the army project convinced us to spinoff Hoana Medical as an independent company to move the technology into traditional acute care hospitals. We decided on the LifeBed as the first product and followed the traditional business practice of interviewing candidate customers—in this case, the nurses tasked with monitoring the patients. We surveyed some 800 nurses and tailored the development of the LifeBed to their responses.

The result was a disaster.

It turns out that what the nurses said they wanted on a survey was different from what they actually needed in a chaotic hospital environment. The procedures they claimed they followed, in a detached interview, didn't account for the shortcuts they took in a life-or-death situation. They didn't have time to drill through multiple menus. Adrenaline kicks in when a patient is dying. When patients were distressed by the device talking, rather than follow a complex shutdown protocol nurses ripped out the plug to silence the machine.

Our engineering team complained that the nurses weren't smart. They didn't read the manual. They didn't get sufficient training. They didn't pay attention. They couldn't use the equipment because they didn't understand it. But that's completely wrong.

The nurses had a job to do, and we weren't helping them do it. We didn't anticipate how they would react.

Any product that needs a manual to work is broken.
—Elon Musk

We'd talked to nurses, but that wasn't sufficient. We hadn't gone through the empathy process, so we didn't really understand them. We hadn't watched them at work.

We had installed LifeBeds on a few nursing floors in a hospital in Indiana. I went there to investigate and observe. I got a nurse to draw me a picture of what she wanted and gave it to the team. "Here's the new design, guys." It had one button.

Henry Ford is credited with many famous quotes, some which are not substantiated but attributed to Mr. Ford nevertheless. One version attributed to him is particularly relevant: "If I'd asked a farmer what he wanted, he would have said a faster horse." One could debate the differences between a car and a horse, but that's not the point. What he is channeling with this story is, to get market adoption of a disruptive technology, you can't use an engineer's perspective, you have to adopt the user's perspective—in this case, the farmer's perspective.

Our experience with the LifeBed is ultimately what got us interested in Design Thinking. We knew there had to be a better way. We *thought* we knew how to design products for users, but we realized we had a lot of room for improvement. In the years since, as noted earlier, Oceanit has embraced Design Thinking to the extent that we now cohost an annual Design Thinking Bootcamp in Hawaii (the first two years with the Stanford team), and we regularly collaborate with the Stanford d.school (School of Design). It's an essential part of our process. The more disruptive a technology is, the more critical Design Thinking becomes.

As you already know from the early chapters of this book, we maintain a healthy skepticism of expertise at Oceanit. Experts can be short sighted, dismissive of new ideas, and premature in declaring an idea or approach unworkable. But because the medical device space was new to us, we felt we had to rely on experts.

Early on we ran into an issue where the team of traditional medical device engineers wanted to run a cable from the sensors embedded in the mattress coverlet to an external box for signal processing—all but one guy, a former Cisco engineer, who asked, "Why aren't we doing this wirelessly?"

The experts all said this is how medical is done, you have to have a cable or the FDA will never allow it. But that cable needed special shielding to control for noise from other hospital gear and would have cost $1,000. The box would cost even more. Plus, it was another part that could wear out and eventually break. Added to this was the irony of a wireless medical sensor system that depended on a cable.

Experts have their place, but they're most effective in industries that are changing slowly and less useful with technology that is changing rapidly. We needed to be more like Cisco and less like HP. In the end, we kept the Cisco engineer and let the others go. And we got our FDA approval. This enabled Hoana to drive down the cost of goods sold (COGS) from about $2,500 to under $200.

Unfortunately, the 2008 recession and the passage of the Affordable Care Act created a lot of uncertainty around medical spending, and Hoana Medical is in the process of regrouping. At its peak, the LifeBed was in thirty-four hospitals, protecting 100,000 patients for over 10 million hours, and in that time performing 4,000 life-saving interventions. Since then, Hoana has continued to improve the technology and reduce the costs, allowing even more sophisticated and widespread applications. The cost of goods for the LifeBed was very expensive, around $2,500. Hoana continues to drive the COGS down and has licensed the technology to Faurecia, an automotive company that wants to outfit the cockpits of semiautonomous cars to monitor the driver's ability to take over. From the data, they can determine if the driver is sleepy, happy, or afraid. It's a radical new possibility for the same tech, but we're still pushing to improve it, to push the cost down another order of magnitude to $20 and open up even more areas of application.

Hoana was the first example of Intelligent Medical Vigilance— automatically observing and analyzing patient conditions and, as necessary, notifying medical personnel. This technology saves lives and reduces health care costs. The big innovation, though, is the passive nature of the oversight. Just as airbags in cars save lives with no additional action required on the part of the driver, so too can Hoana technology, which will eventually find its way into everything from automobile seats to airline cockpits.

The health-monitoring apps on the Apple Watch, which have already been credited with saving lives, are a consumer-oriented version of the same idea.

////////////

Implications

At Oceanit, we put great thought into the design of our products so that people will want to use them. The most advanced expression of that is technology that people don't even have to be conscious of—it just exists and does its work.

The truly amazing technology of the contactless medical sensors is the result of the deep science of digital signal processing (DSP), new materials, low-cost manufacturing, and a host of other engineering breakthroughs that have all converged to enable this technology to be highly reliable and very inexpensive. However, how this technology can ultimately be used more broadly to impact humans and society is still a chapter being written.

As the world's population ages and the cost of care becomes challenging, this type of technology could change the calculus of health care delivery. A contactless medical sensor at home could provide a near continuous checkup, alerting patients of emergency health challenges before they become acute. No compliance is required of the patient, since the system is totally invisible and unnoticeable. Rather than hoping patients will understand and comply with a complicated discharge instruction booklet, after-care and maintenance for many processes will be instantaneous when needed. If there is an adverse event, people will be notified immediately. It's the kind of thing you would see in *Star Trek*, but it's available and very affordable now. This is how we drive down the cost and improve the quality of health care delivery. Future implications of such technology would include the following:

- *Aging in place housing.* Rather than move to assisted living, there are many people who can age in place if adequate support and safety is available. Emerging sensing technology is just beyond the horizon and should be available and economically accessible within in the next decade.

- *Cockpit monitoring and safety.* Whether piloting a car, an aircraft, or other large piece of machinery, sensing in the cockpit to track the condition of the pilot will reduce the risk of pilot fatigue, accidentally falling asleep, or other challenges that can lead to deadly accidents. This will be particularly important as artificial intelligence becomes the standard, where autonomous vehicles will handle most situations but on occasion the pilot will need to be thrust into action to deal with outlier events. To reduce the risk of trusting a half-asleep pilot to take over controls, monitoring their condition of drowsiness will be imperative.

- *General wellness and the continuous checkup.* Although there are many devices that personally monitor one's heartbeat, number of steps, and so forth, the jury is still out on the final configuration and the ultimate noncompliant utilization of these technologies. When this occurs, there will be an explosive market opportunity to improve one's overall wellness, reducing the overall cost of healthcare, and improving the quality of life.

With all the talk in prior chapters about culture, one of the first things I need to acknowledge is that to transition from science to products is difficult because the cultures of science and human-centered design are from different worlds. We bridge the culture between these two worlds by using Design Thinking, not to develop the design per se but to establish a common language, since we believe language is key to bridging cultures. Developing the breakthrough science is only half the battle. The other half is making the resulting technology available and usable.

We have a phenomenal science and engineering research establishment across the United States. At least 225 engineering schools offer the PhD—places like MIT, Caltech, Stanford, Rensselaer Polytechnic Institute, Rice. Then add to this picture the seventeen national labs—such as John's Hopkins Applied Physics Lab, Georgia Tech, Lawrence Livermore, plus the several dozen other Federally Funded Research and Development Centers (FFRDCs).

As an example, I recently visited the Physics Department and the College of Engineering at the University of Colorado in Boulder. UC Boulder is a public university, but it continues to push the edges of science and engineering. With one of the largest physics programs in the United States, it has produced five Nobel laureates, four in Physics. The College of Engineering continues to grow, with the upgraded Department of Aerospace Engineering bringing on line a new state-of-the-art research complex that takes your breath away, supporting 1,200 students and hundreds of PhD students.

Unlocking the amazing science and discoveries occurring in US universities and national laboratories, as well as in companies that are willing to dive into the most difficult problems, has profound implications. However, most of these discoveries won't see the light of day. As I described earlier, they isolate themselves in the Blue Zone, the world of deep science and discovery. Then you also have some amazing product design companies. IDEO, one that we've followed over the years, became the inspiration for the Stanford d.school. They live in the Green Zone, producing products to impact humans and society. If there ever was a way to consistently connect the Blue and Green Zones, the world would change in ways beyond description. Welcome to the Rock & Roll Zone, where we use a common language to connect two very different cultures. Just as the kinetics of business are driving economic change, the national and global resource infrastructure is always expanding as the edge of science continues to get pushed out at an accelerated pace. Connecting what science and technology can provide with what society needs is the key to a better world, not least by

- controlling climate change;
- improving health and longevity; and
- spreading economic prosperity.

The key is as simple as it is difficult: how to enable communication and mutual respect between those who live in the Blue Zone and the Green Zone. I've offered one model for how this can work. No doubt there are others. It's surely worth continued exploration and experimentation.

CHAPTER 12

Mind-to-Market:
Delivering Innovation into the Hands of Users

We're doing the grinding, sometimes frustrating work
of delivering change—inch by inch, day by day.
—Barack Obama

In 2012, the once proud Eastman Kodak Company, the firm that dominated still and motion photography for over a century, filed for Chapter 11 bankruptcy protection. The rise of digital imaging had slowly hollowed out their market until they were forced to sell off their photographic film business and reorganize—another dinosaur disrupted by quicker competitors.

Yet it was Kodak itself that had invented the digital camera more than thirty years earlier. Kodak was at the absolute forefront of the digital imaging revolution, but they let the technology languish in the lab in order to protect their film stock and processing businesses—the same businesses they were forced to unload in bankruptcy reorganization. They squandered a decades-long lead because they didn't have a strategy for disruptive innovation.

Disruptive innovation fails to reach the market for any number of reasons: fear of self-disruption, as with Kodak; an internal culture clash; an inability to scale; or a lack of resources, particularly financing. It's important to have a strategy for bringing innovation to market.

PATRICK K. SULLIVAN

The more disruptive an idea or technology is, the more important it is to understand the path to market. That's the concept of Mind-to-Market, a systematic method for shepherding breakthroughs all the way from the lab into the hands of paying customers.

There are many risks around bringing new technology to market: technical risk, market risk, and execution risk. Any one of these is enough to scuttle a new technology. The bulk of this book is dedicated to technical risk, the uncertainty around developing new technology, and specifically to addressing technical risk through the Intellectual Anarchy process. This chapter focuses on reducing market and execution risk through various go-to-market modalities. To bring innovation successfully to market, we must innovate our go-to-market strategies as well as the technology itself.

Start-up Financing

For most start-ups, financing risk becomes the all-consuming issue, as the new company has to raise capital. When successfully executed, the capital raised is directed to specific milestones that de-risk the opportunity, on the assumption that additional capital will follow once that milestone is reached.

With each round of financing, investment capital takes an additional share of company ownership, typically 20–30 percent per round. After a Series A, Series B, Series C, and the sale of preferred stock, the founders are lucky to be left with a single-digit percentage of their own company.

This trade-off can be worth it in the case of a company with a multibillion-dollar valuation, but those are so rare that they're referred to as "Unicorns."

Investor exits typically occur through an acquisition or an initial public offering (IPO), which enables the company to trade stock on an exchange.

	Series A Dec-02 $0.34			Series B Dec-03 $0.45			Series C Nov-05 $0.60			Series D Jul-06 $1.43		
Price per share													
Investors	**# Shares**	**$**	**% O/S**	**Shares**	**$**	**% O/S**	**Shares**	**$**	**% O/S**	**Shares**	**$**	**% O/S**
Common Shareholders	9,100,000		46.0%	9,100,000		28.5%	9,100,000		18.3%	9,298,699		13.3%
Employee Option Pool	2,900,000		14.7%	3,900,000		12.2%	4,900,000		9.8%	7,201,301		10.3%
Common Share Warrants	200,000		1.0%	200,000		0.6%	200,000		0.4%	488,888		0.7%
Series A Investors	7,573,529	$2.58M	38.3%	7,573,529		23.8%	7,573,529		15.2%	7,573,529		10.8%
Series B Investors				11,111,111	$5.00M	34.8%	11,111,111		22.3%	11,111,111		15.8%
Series C Investors							16,958,320	$10.17M	34.0%	16,958,320		24.2%
Series D Investors										17,482,517	$25.00M	24.9%
Total	19,773,529	$2.58M	100%	31,884,640	$5.00M	100%	49,842,960	$10.17M	100%	70,114,365	$25.00M	100%
Money raised	$2.58M			$5.00M			$10.17M			$25.00M		
Pre-money valuation	$4.15M			$9.35M			$19.73M			$75.26M		
Post-money valuation	$6.72M			$14.35M			$29.91M			$100.26M		
Milestones	1) Recruit Mgmt Team			1) Preclinical Data			1) Complete 1st Product			1) Sales Team Expansion		
	2) Build Dev Platform - Alpha3			2) Preproduction Prototype			2) Obtain 510(k) Clearance			2) Enter Second Market		
	3) Identify 1st Market			3) Customer Income			3) Begin Sales			3) Positive Cash Flow		

NOTE: *Investment amounts are in millions (M) of USD and rounded for illustration purposes only.*

Fig. 12.1. Sample capitalization table that shows multiple rounds of financing: Series A to Series D preferred stock sales used to raise capital. With each round, there is commensurate equity dilution, but this can be offset by increasing value—market capitalization value. Investment numbers shown are rounded to simply show the impact of additional equity capital.

INVESTOR RETURNS:

Year of Liquidity Event:	2008
Market Cap:	$869.55M
Price per Share:	$12.40

Shareholders	Shares	Ownership	Investment	Liquidation Value	Multiple
Common Shareholders	9,298,699	13%		$115.32M	
Employee Option Pool	7,201,301	10%		$89.31M	
Common Share Warrants	488,888	1%		$6.10M	
Series A Investors	7,573,529	11%	$2.58M	$93.93M	36
Series B Investors	11,111,111	16%	$5.00M	$137.80M	28
Series C Investors	16,958,320	24%	$10.17M	$210.32M	21
Series D Investors	17,482,517	25%	$25.00M	$216.82M	9
Total	70,114,365	100.0%	($27,575,000)	$869.55M	

IRR:	2002	2003	2004	2005	2006	2007	2008	IRR
Series A Investment	$2.58M	$0	$0	$0	$0	$0	$93.93M	82%
Series B Investment	$0	$5.00M	$0	$0	$0	$0	$137.80M	94%
Series C Investment	$0	$0	$0	$10.17M	$0	$0	$210.32M	174%
Series D Investment	$0	$0	$0	$0	$25.00M	$0	$216.82M	194%

NOTE: Investment amounts are in millions (M) of USD and rounded for illustration purposes.

Fig. 12.2. Sample investor returns show the relationship between risk and reward for high-risk investors. Venture returns need to be high to offset venture-investing risk, where most start-ups fail in the first few years. However, when managed in a portfolio, one success can offset all the losses.

In the early stages of a new company, particularly start-ups, execution risk is dominated by financing risk. Raising capital becomes the all-consuming issue. The urgency of securing a Series A round, or the next milestone, can blind founders to technical and market risks and consume resources better spent elsewhere. But the most common failure mode for start-ups is market timing.

Bill Gross of Idealab reviewed start-up failures and found that roughly 40 percent of all start-ups fail because they are either too early to market or too late. Companies that fail to connect with the market need additional financing to sustain themselves, but this is when capital becomes the most scarce and expensive. Hence, what seems to be financing risk, a failure to raise capital when needed, could actually be solved by addressing market risk. This is precisely the solution we arrived at with STAMPS (Strategic Technology Acquisition Managed Process), discussed later in this chapter.

Declining Early Stage Medical Investment

Each market sector has its own go-to-market profile. A medical device takes seven or eight years of development and a $50 million capital investment, while a therapeutic requires ten to twelve years and $150 million. Any significant missteps require additional capital. Meanwhile, a phone app with the potential to become a Unicorn can be funded with very little capital.

With the changes and uncertainty in the US health care policy, it's no wonder that investment in early stage medical device and therapeutic companies has largely dried up.

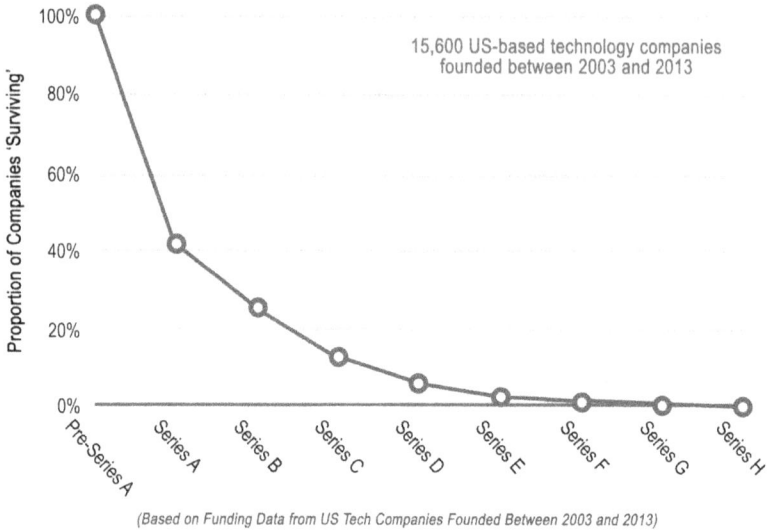

100% — 15,600 US-based technology companies founded between 2003 and 2013

Fig. 12.3. Statistically, 90 percent of start-ups fail. Unfortunately, the start-up culture celebrates raising Pre-Series A or Series A financing as though that's the end game. However, that's just the start of a relentless search for further financing, where follow-on financing drops off precipitously. (Crunchbase)

Venture Capital

Currently venture capital (VC) dominates the start-up funding landscape, both in terms of dollars invested and mindshare. If you talk to university students, business people, or researchers, there's a belief that venture funding is almost synonymous with successful innovation. There's been a trend toward start-up incubators recently, but even these are merely springboards for eventual venture financing.

There's no question that venture capital has been an amazing innovation in itself, helping to bring into existence tech giants like Facebook, Uber, and Google. However, venture capital is almost unavailable outside its areas of concentration. In the United States, the bulk of venture capital has accumulated in just a few regions—New York, Boston, Southern California, and the San Francisco Bay Area.

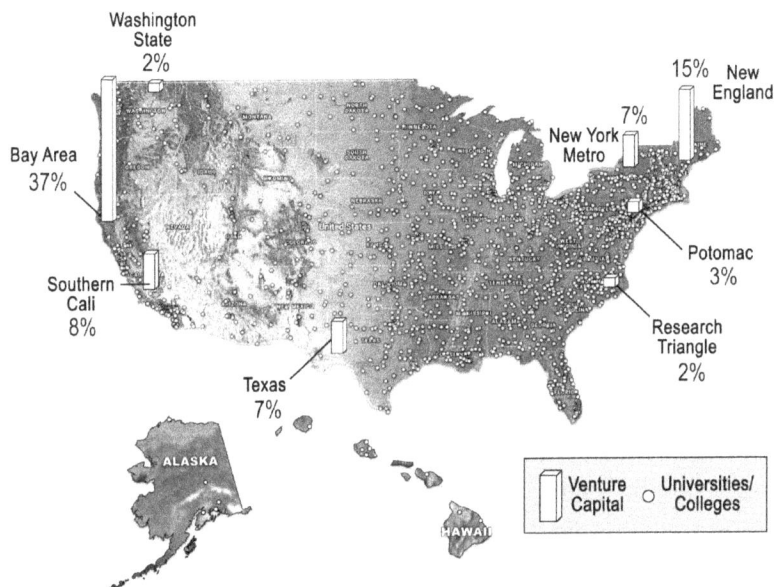

Fig. 12.4. There are 5,000–6,000 colleges and universities distributed across the United States that offer STEM education. However, venture capital for start-ups is limited to eight locations, principally four, and is practically unavailable to the rest of the country. As a rule of thumb, if you are outside a thirty-minute drive from the venture capitalist's office, your chances of raising capital effectively disappear. As a result, most classically trained start-ups must migrate to just a few locations around the United States.

This geographic clustering is an artifact of the way venture funds are managed. The vast majority of individual venture-backed firms fail, but the few that do succeed perform so spectacularly that they more than make up for the losses. To ensure that they score a winner in their portfolio, venture capitalists need access to a lot of deals, and they do those deals with nearby companies so they can oversee and assist the companies they've invested in. As a result, they tend to limit their investments to companies within a thirty-minute drive from their office. Capital becomes even more concentrated as start-ups chasing funding move closer to the money, and more deal-hungry funds appear to finance those start-ups.

Risk and the Venture Capital Model

The venture capital business works on the "two and twenty" model—a 2 percent annual management fee plus a 20 percent carry (performance) fee. Thus, a billion-dollar fund collects a $20 million management fee per year, regardless of success or failure, plus a 20 percent share of the profits, if any. Because individual deals are so high risk—most venture-backed firms fail within a few years—funds are typically structured to last ten years and invest in many deals to ensure a positive return.

One would think that this "free roll" model would lead to incredibly risky behavior. Imagine if a friend gave you $10,000 to wager in Las Vegas under the same conditions. You could bet on Red, for a roughly 50-50 chance at $2,000 (your cut), or you could lay it all on your favorite number. Your chances shrink to 1 in 38, but if you hit, you stand to make $70,000 (your friend makes $280,000). Remember, it costs you nothing if you lose, and either way you pocket $200 for your trouble.

However, venture capitalists have become extremely conservative in their investments, coming in at as late a stage as possible and only after a company has eliminated most risk by demonstrating market traction. Even then, they prefer to join in only if another venture capitalist is already on board, resulting in a pack mentality that has resulted in some spectacular failures (Juicero) and is generally bad for innovation.

Venture capital tends to spring up around major research universities. Building a company requires a lot of energy and hard work that young, well-educated college students are willing to provide. There are about 150 great research universities in the United States but nearly 5,000 that provide a nearly equivalent STEM education. This leads states to try to mimic Silicon Valley, only to discover that there is only one Silicon Valley. The network effects of concentrated capital are too powerful to overcome.

I'd argue that the ultimate innovation of Silicon Valley has been the finance and monetization model that made the current wave of technology companies possible, even more than the companies themselves. But that very venture capital model has now brought innovation to a halt. There's very little discussion around business and finance innovation now. We need more.

While certain schools are superior, I would argue that the basic STEM education is available in any college or university. Venture capital lives in only perhaps eight places in the United States, yet most US colleges and universities teach a start-up model associated with venture capital that is unavailable in their region. As a result, young people are drawn away, and communities are robbed of local talent and energy. To bring economic prosperity to the entire country, new innovations for financing start-ups are needed. Each state should evaluate how it invests capital for retirement systems and pensions. If a fraction of their alternative asset class were available for regional start-ups, their entire community would prosper.

Innovating Technology Financing

If you are like most people in the United States, you don't live within thirty minutes of significant venture capital, so there is essentially no access to capital. Unless you move closer to where venture capital lives, you are out of luck—unless you use a more self-financing approach we refer to as Corporate Co-Development. However, to get to corporate capital, there are many sources available, including the National Science Foundation (NSF), the Department of Defense (DoD), plus another nine federal agencies that all publish solicitations for interesting problems through what they refer to as Broad Agency Announcements (BAA), State of Qualifications (SOQ), and an alphabet soup of acronyms—essentially representing government agencies reaching out for help in solving a technical problem or addressing an unaddressed requirement.

One approach that is interesting for small businesses is the Small Business Innovation Research (SBIR) program, developed by Dr. Rolland Tibbetts at NSF in 1977. I had the opportunity to spend time with Rolland many years ago. He was convinced that the most innovative ideas come from small groups of people that bring creativity, education, and agility together to invent new technology. Today this program has become a source of risk capital available to the entire United States, amounting to close to $2 billion annually. When combined with other sources of capital, one can develop technology without access to venture capital, as shown in figure 12.5.

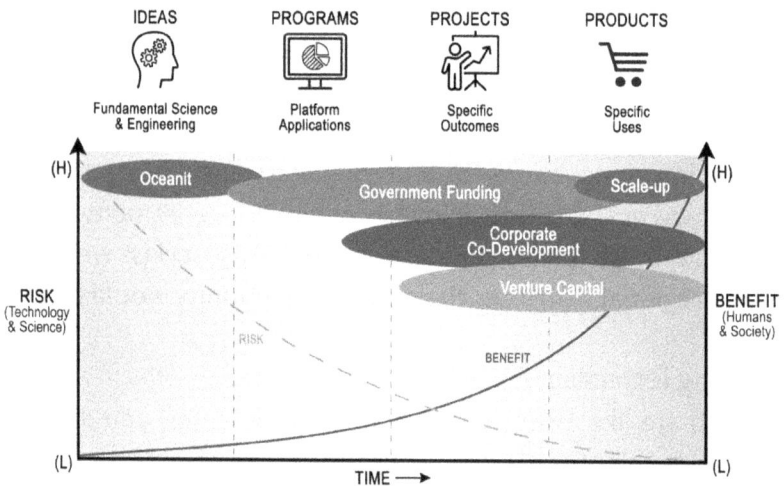

Fig. 12.5. We have a process of reducing risk and increasing benefit with funding from multiple sources. Moving from our home-grown ideas, we then develop and sell to research programs that become projects and eventually products.

We take on the initial "Ideas" risk, typically deciding on an interesting and important "big idea," in quarter four of each year. We also take on other ideas with our "Innovation Fund" twice per year. We then use "government funding" to reduce the risk. Thereafter, we access "Corporate Co-Development" partners to further reduce risk. At this point, we have the option of going to a venture investor or other modalities of going to market, as discussed below. What we are finding is that Venture Capital is moving to the right (less risk appetite) and Corporate Co-Development partners are moving to the left (greater risk appetite) so they can better fill their technology pipeline with new opportunities.

An example of how this works is illustrated in figure 12.6 and is also discussed in more detail later in the book. However, it's a good example to see how we disintermediate Venture Capital and take early risk on ideas that may not yet have a requirement. In this case, the "Interesting and Important" nanotechnology projects, with additional funding from "Programs" and "Projects," significantly reduced the risk of our ideas, while simultaneously revealing a plethora of benefits. Corporate Co-Development partners became very interested in collaborating, providing additional capital that eventually produced a broad portfolio of products— probably close to thirty nanotechnology product opportunities.

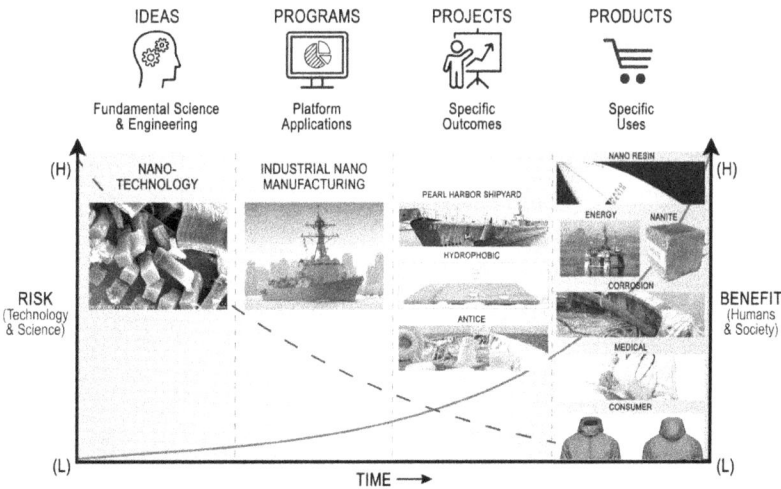

Fig. 12.6. An example of nontraditional financing that disintermediates venture capital, which translated big ideas in fundamental nanotechnology science into a portfolio of products. We started with the "interesting and important" promise of nanotechnology, which eventually delivered nearly 30 nanotech products across numerous industries.

Licensing

An alternative go-to-market strategy is basic licensing, which seems straightforward—you grant a company the right to use your technology or intellectual property (IP) in exchange for a percentage or per-unit fee. The reality is more complicated. Most licensing deals with universities and national labs don't amount to much because of the challenges of getting to scale. It's one thing to create a proof of concept in the controlled environment of the lab, and yet another to deal with the complexities of the real world. Scaling up is hard work and difficult to do well, and most lab researchers aren't interested in the tedious, repetitive detailed work of bringing something to market. It requires a completely different skill set. We've learned that people who thrive in open-ended research and discovery will slowly die in the hyperfocused environment of perfecting a product for the market.

In terms of risk, the licensor is responsible for the technical risk, and the licensee assumes the burden of market and execution risk. Therefore, most licensing fails to deliver real value, since the challenges of market and execution risk are not insignificant.

Additionally, the population of potential licensees is small. Perhaps one or two companies in the world might truly understand the opportunity and be motivated enough to put in the hard work and significant investment required to bring the technology to market. Finding these businesses is a needle-in-the-haystack problem.

Acquisition

The challenges of licensing predispose most businesses to favor acquiring companies instead. An acquisition can significantly reduce execution risk, as the employees who developed the technology come along as part of the package. Typically, the acquiring company already has market distribution or can add the product into their sales channel using their existing sales force. The downside is that most employees who come with an acquisition will leave after their one- or two-year lockup period expires. In addition to the headaches of employee integration, culture clash, and organizational change, the acquiring company ultimately ends up having to train its own workforce.

From a pure market standpoint, however, there are more potential acquirers than licensors, since acquirers can be either from related or totally unrelated enterprises. Investment bankers love acquisitions; they collect a 5–10 percent fee on the transaction, and a banker workforce can spread out and make inquiries with a large pool of candidate acquirers. By the time a company is in the market, selling product to customers, it's pretty easy to put together a seller's prospectus, like a realtor selling a home.

Strategic Technology Acquisition Managed Process (STAMP)

At Oceanit we've been working on a new category that we call STAMP—Strategic Technology Acquisition Managed Process— that addresses the limitations of licensing and acquisition. Unlike licensing, a STAMP allows you access to technology that is prescaled and field tested. Unlike an acquisition, a STAMP allows you to essentially buy the company without acquiring the people. We provide a temporary workforce for technology onboarding that returns to Oceanit at the end of the process, leaving the acquirer with its own trained workforce.

	Pure IP Acquisition/Lic.	STAMP Acquisition	EBITDA Acquisition
Science	✓	✓	✓
Proof of Concept	✓	✓	✓
Small Scale Demo		✓	✓
Field Demo		✓	✓
Industrial Scale-Up		✓	✓
Field Deployment		✓	✓
Customer Demand		✓	✓
Technology Onboarding		✓	✓
EBITA Growth			✓
Employee Integration			✓
Organizational Change			

Fig. 12.7. Features of a STAMP: Strategic Technology Acquisition Managed Process compared to traditional licensing or EBITA acquisition.

A STAMP can be a Corporate Co-Development relationship that resembles an acquisition. With a STAMP, Oceanit assumes the technology risk. The market risk is mitigated because the corporate partner essentially *is* the market, so market fit and market timing are nonissues. We manage execution risk jointly through the Corporate Co-Development relationship, where the corporate partner puts up nonrecurring engineering (NRE), and Oceanit provides the know-how and manpower to perfect the product to their exact needs and scale to address market demands.

This approach benefits both parties. As a STAMP grantor, we gain streamlined access to a ready market. The STAMP acquirer gains proven technology tailored to their specific requirements, along with assistance in integrating it into their processes, supply chains, quality systems, and so on. Occasionally we help with organization change management (OCM) or process reengineering as well.

Most acquirers don't have the capacity to adequately integrate an acquisition, let alone the new technology that comes with it. That's why we provide eighteen to twenty-four months of hand-holding to ensure the STAMP is successful.

A STAMP, together with the Corporate Co-Development financing, effectively disintermediates venture capital by developing a product that can be acquired by companies and inserted into their portfolio of offerings, which also includes technology training, onboarding, training, and quality systems. The STAMP offers a new product for investment bankers, in addition to mergers and acquisitions (M&A) or initial public offerings (IPO's). The optimal time for both technology creators and acquirers is just past "growth demand." The tradeoff is to either raise outside capital, which is highly dilutive as well as major distraction, or sell to an acquirer, where there is no dilution but less value, since the company is not a full EBITDA acquisition target.

The calculus of finding the intersection of dilution and discounted value, all tempered against market risk, is where pricing occurs. This works well for strategic acquirers, since traditional M&A's typically fail to effectively integrate into the acquiring company. The downside is that you give up the possibility of the grand slam you might achieve with venture capital, but those outcomes are rare. The benefit is that the entire process moves much faster than a venture-financed deal, where you need to recruit a CEO, assist with financing, and invest significant time managing incidental details, and you are dependent on the maturity of your team and the experience of the venture investors. There is a good reason venture investors refuse to coinvest with certain firms: life is too short to deal with certain types of investors. The STAMP option reduces the risk of "difficult" investors.

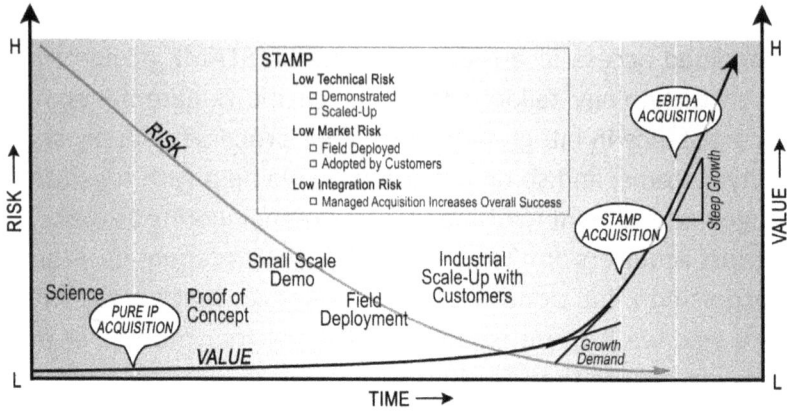

Fig. 12.8. Risk of acquisition goes down as the technology gets closer to being proven in the market. Moving into the market with customers also increases value because it validates market acceptance.

Because of their agility and market specificity, STAMPs are particularly effective for addressing hidden requirements. We find that companies often don't make their actual market requirements and specific details available publicly. They don't want to talk about what's not working. It's typical for a Fortune 100 company to post a list of "important challenges" for their organization, but when we partner with these companies, the issues we end up working on are almost never on that list. This is true for just about every large organization we work with. You need to get deep into the bowels of an organization to understand the issues they're really grappling with, as opposed to the ones they're willing to admit publicly.

Doing this requires building trust, understanding, and empathy, but by working with their subject matter experts (SMEs), we can quickly get to the details of the problem. These details tend to be incredibly specific, since the issue may have been locked away inside their corporate research arm, unsolved, for decades. We're able to draw on our thirty-plus years of innovative research and solutions—some mature, some for which there is no current market—and rapidly deliver a solution.

238

This agility enables us to partner with established companies and enter brand-new and emerging markets. For example, when we began research into nanoenergetics, we were merely curious about the potential for controlling combustion on a nanoscale. The influx of the private sector into the space industry, particularly with microsatellites, has opened up a new market for our work in nanopropellants that we never could have predicted.

Portfolio Management

Although getting to market with a STAMP can be relatively fast once we've found the right corporate partner to work with, the long process of discovery and exploration leading up to it can represent an investment of a decade or more and tens of millions of dollars. Amortizing that cost requires a version of portfolio management theory.

It's my belief that risk-adjusted portfolio management is the future of Mind-to-Market, allowing companies with long time horizons to effectively mitigate technical, market, and execution risk themselves.

When you rely on a venture capitalist, *they* gain the benefits of a diverse portfolio. They invest in dozens of companies alongside yours, sometimes in the same market, expecting most of them to fail. They're protected by making many simultaneous bets. However, with no risk mitigation from portfolio theory, when all your money is in one company, you succeed or fail as a single event. Keep in mind that the vast majority of venture-backed firms do fail.

By relying on STAMPs, you gain the advantages of a venture firm for yourself, bringing your portfolio in-house. Significantly speeding up the financing process and reducing the market risk enables you to make many small bets instead of one big one. It's a genuine innovation in the economics of research and essentially the inverse of the venture capital model. You give up the opportunity for a Unicorn, but effective risk-adjusted portfolio management enables more predictable outcomes in bringing the deep science discoveries to market.

I discuss the technical aspects of the long pipeline of discovery and innovation in much more detail in the final chapter of this book, "Left of Boom and the Intellectual Anarchy Process."

The First Thousand Feet: Secure Drilling with Steel-Cement Interfacial Nanobond

Anything you build on a large scale
or with intense passion invites chaos.
—Francis Ford Coppola

I recently had the opportunity to attend the Offshore Technology Conference (OTC) in Houston, the world's biggest oil and gas conference. I was there to present on new technology, meet new partners, and develop existing partnerships.

I am always stunned at the physical scale of the industry. It is, as author Douglas Adams might have said, "big. Really big. You just won't believe how vastly, hugely, mind-bogglingly big it is." I saw machines the size of houses, and others the size of multi-story buildings. There were seemingly ordinary parts blown up to massive proportions: ropes 3 feet in diameter, threaded bolts too heavy to carry without a forklift, and shackles as large as my kitchen. The sheer size of the equipment on display made me feel as if I'd been shrunk in a science-fiction movie.

Everything about the industry operates on a superhuman scale. Exploratory drilling now reaches up to 7 miles below the surface. If you were to lay the pipe flat, it would take you about an hour to walk from end to end. Successful wells can pump hundreds of barrels per hour and thousands of barrels per day. Four-hundred-meter-long supertankers the size of skyscrapers carry millions of barrels in their holds. The world's total output of crude oil is a staggering 100 million barrels per day.

And then there's the money.

At today's prices, the total value of the oil and gas industry is more than $2.5 *trillion*, roughly 5 percent of the world's economy.

But each drop or rise in price shifts the economics of the industry. As recently as 2014, oil was over $100 per barrel. A few years later it fell below $35. Today it hovers around $50.

The major players can be divided into two categories: the oil companies and the service providers. They work in partnership, but their interests aren't perfectly aligned. The former include the international oil companies (IOCs), such as Royal Dutch Shell, Chevron, Exxon, British Petroleum (BP), and the national oil companies (NOCs) that are partially or completely state owned, such as Aramco (the Saudi-owned Arabian American Company), Petrobras (the Brazilian national company), PEMEX in Mexico, and so on. The IOCs and NOCs are responsible for exploration and production. They bear the brunt of the risk and reap the bulk of the profit—when there is any.

The service providers, on the other hand, operate largely free of risk. They get paid either way. These providers include companies such as Schlumberger, Baker-Hughes, and Halliburton, which provide downhole services, equipment, and cement.

The unequal distribution of risk puts the oil companies at odds with the service providers. The oil companies are naturally inclined to push the schedule as hard as possible to move the well to production and profitability. The service providers are incentivized to stall in the name of safety. If the job takes longer, they get paid more. You can see this dynamic play out in the movie *Deepwater Horizon* about the Macondo oil blowout in the Gulf of Mexico.

The film opens with a debate over safety. Mark Wahlberg's character (Chief Electronics Technician Michael Williams) arrives to begin his extended shift on the rig and learns that the workers responsible for pouring the cement to secure the steel casing have left without conducting a pressure test. This casing, essentially a large steel pipe, must be secured to the surrounding geology to stabilize it and prevent leaks. The first 1,000 feet through rock are critical. In the case of the offshore Macondo well, the pipe stretched down through 5,000 feet of water, and then the next 1,000-foot section had to be bonded to the sea bed with concrete.

Managers from British Petroleum sent the concrete workers back ashore without conducting a pressure test.

In the event of a failed pressure test, a recementing, referred to as a "squeeze job," would be necessary—pumping more cement downhole and forcing it up the exterior sides to secure the pipe to the surrounding geologic formation. A squeeze job for a deep well can range from a few million dollars to tens of millions, depending on the challenges.

Cement vs. Concrete

The terms "cement" and "concrete" are often used interchangeably, but in fact cement is just one ingredient of concrete and of little use on its own. Cement is an adhesive, but concrete is a mix of sand, gravel, and rock—collectively called aggregates—plus cement and water. The cement binds the aggregates together into a solid, stable structure. Without them, the cement would be weak and flimsy.

You can think of cured concrete as a Rice Krispy treat. The melted marshmallows are the adhesive—the cement. The Rice Krispies are the aggregates—the sand, rock, and gravel. Without the Rice Krispies, you'd just have a sticky mess on your plate. With the marshmallows binding the Rice Krispies together, you get a solid brick.

When we're focused on the adhesive properties of the mixture, as I am in this chapter, we'll sometimes prefer the term "cement," knowing that we're still actually talking about concrete. When we refer to a finished product like a sidewalk or the foundation for a house, we're more likely to use the term "concrete."

Faced with costly delays and overruns, the managers from BP, the oil company, skipped this pressure test step so they could move the well into production and begin recouping costs. We know how this story turned out: a massive oil spill, eleven lives lost, and extensive environmental damage. BP has since paid out nearly $65 billion in fines, civil litigation, and cleanup costs. Halliburton, the service provider responsible for the cement, was fined about $200,000, paid over $1 billion in legal settlements, and contributed an additional donation to help offset the impact in the affected communities.

With the decline in oil prices, the risk differential between the oil companies and the service providers is accentuated, increasing tensions. Costs stay high even when oil prices decline, so profit margins are squeezed. As a result, the oil companies have discovered a new appetite for disruptive technology: anything that will allow them to turn a profit in this new era. This is good news for companies that specialize in providing such technology—companies such as Oceanit.

One example is the development of our Steel-Cement Interfacial Nanobond (SCIN), a nanotechnology treatment for the steel casing of a well. We've already seen how the quality of the steel-cement bond is critical to producing a robust, reliable well. If the top 1,000 feet of steel casing are securely connected to the geology, the risk of a catastrophic accident like the Macondo blowout is significantly reduced. The SCIN treatment improves this critical bond by about 100 percent, changing the surface behavior of the steel casing, allowing the cement to chemically bind much more tightly.

This not only creates a safer well but a cheaper well, as fewer expensive squeeze jobs are needed. Where normally ten squeeze jobs might be necessary to secure a pipe, with SCIN that number might drop to two, making the economics of inexpensive oil in tough environments workable. The risks associated with exploration are unchanged, but the cost of production is reduced, bringing the incentives of the oil companies and the service providers into closer alignment. This is a great example of market forces inspiring market leaders to introduce disruptive technology.

When Shell expressed interest in SCIN, we had to go from making it work on a 5-inch coupon sample in our lab to 5,000 feet of steel casing in the Marcellus Shale formation in Pennsylvania in a matter of months.

This scaling-up process presented a number of challenges, from physical and chemical to cultural. Instead of mixing solutions in carefully calibrated beakers a few milliliters at a time, we had to prepare them in batches of hundreds of gallons at a time. And we had to coordinate with industry personnel, who are very different from scientists. They have different skill sets, but they also have different personalities. Getting PhDs to talk with and understand roughnecks and roustabouts was not easy.

In fact, our first attempt at scaling up didn't go well because we didn't have a system in place for quality control at that scale. We were treating 40 to 50-foot sections of steel casing, each of which could weigh a ton or more. Under pressure, we put together a new custom quality control system, a unique mix of high tech and low tech, from a hacked multispectral camera and some ISR (intelligence, surveillance, and reconnaissance) software we'd built for real-time military video analytics, all mounted on an ordinary skateboard.

The new system could be quickly and easily skated across the surface of the pipe, producing a thorough inspection report. This enabled us to iterate the scale-up process to produce a uniform, nanotreated casing.

And this enabled us to deploy into some of the most aggressive wells in the Gulf of Mexico.

This is the arena where so many promising disruptive technologies fail, in the gap between the potential and the actual—not because they can't be scaled up or moved out of the lab, but because the disconnect between the two environments is so extreme. The scientists who develop the technology and understand its potential don't have the skills or the inclination to push it out the door. And those who have the manufacturing and distributing capability either fail to recognize the potential or are protecting entrenched interests.

In our case, we managed the scaling-up process effectively, and Shell has been deploying SCIN into wells in the Gulf of Mexico.

///////////

Implications

In drilling, whether it's for fossil fuels or drinking water, you have to start with steel meeting geology, or at least its human-made version, concrete—pretty simple, solid elements, right? But much can go wrong at this junction.

SCIN is a great example of how industrial nanotechnology can transform a technology that's been around for about a century, enabling what was considered impossible to become routine.

In the energy space, the biggest benefit of SCIN technology is that it improves the bond strength of the cement bond between a well and the geology by about 100 percent. This significantly upgrades the safety, quality, and reliability of oil and gas reserves. SCIN enables higher "zonal isolation," whereby materials recovered from the depths are much more likely to stay in the pipe and not mix with groundwater and other environments on the way up the recovery infrastructure. With SCIN technology, accidents like the Macondo Well explosion in the Gulf of Mexico perhaps would not have happened.

The world is running out of clean water. SCIN could improve water well integrity and reliability while reducing costs. This would have a lifesaving impact in locations where the budget and expertise for safe water system maintenance is unavailable.

Longer term, SCIN changes the way we think about materials and composites, particularly by improving the marriage of cement and steel. SCIN–like treatments can have significant impact to industries we all rely on. Here are some examples:

- *Durable composites.* This would include everything from nuclear energy shielding to undersea vehicles, where external stresses can cause the different materials to expand or contract differently, producing stress at the joints.

- *Paint adhesion for high-stress environments.* Relying on improvements in adhesion from paint will open up new markets for specialty paints specifically needed in industrial processes.
- *Reconstructive surgery.* New capabilities in surface adhesion—everything from dentistry, where ceramic caps can be added, to repair damaged or cracked teeth, to bone grafting techniques used by orthopedic surgeons—could improve from new capabilities in surface adhesion.

/////////

The more disruptive the technology, the more important it is to understand its path to market. It's not about geography, it's about an educated workforce, willing to innovate and collaborate. Oceanit's Mind-to-Market model is a systematic method for shepherding breakthroughs all the way from the lab and into the hands of paying customers. Just as venture capital has been an extraordinary innovation that enables technology to go to market, there are other ways to accomplish the same outcome. An interesting review of approaches to financing are discussed by Glenn Yago in his book, *Financing the Future,* and *Financial Innovation*, by Glenn Yago et al., where he chronicles innovation in finance. Another striking model is how the Israelis developed their tech industry, as described in the book *Startup Nation* by Dan Senor and Saul Singer.

If you have a background in business, you might remember a class on the marketing mix, which comes close to Mind-to-Market. The "4 Ps" were taught as Price, Product, Promotion, and Place. Place really means Distribution, but no one wants to remember three Ps and a D. Put them all together and you have, theoretically, a winning product. Mind-to-Market is like that, on steroids, driving innovation not just in the product, but in how affordably it can be made and how directly and immediately it can be put in the hands of customers. Add a mindset of nonlinearity and business agility and you are gaining a good understanding of how we operate. We don't try to apply any step-by-step instructions for innovating in an unknown space. We can pivot whenever necessary.

Mind-to-Market requires a willingness and determination to innovate through the entire lifecycle, from ideation to experimentation, from financing to scaling up, to partnerships and market distribution. The Mind-to-Market approach can use different modalities to deliver solutions to the particular customer. In the case of SCIN, it wasn't enough to innovate the science and technology of the product. We needed to adapt to the financial culture that drives the oil and gas market. We accomplished this by creating a specifically crafted Corporate Co-Development model that effectively disintermediated venture capital, while providing the financing needed. In a nutshell, we brought the technology but perfected it for specific market requirements. It was an excellent partnership that produced impressive results with all the agility of a start-up, but with the specific market driven challenges of a Fortune 10 company.

When we started this project, we were told that other groups had worked on how to improve the cement-steel interface for more than two decades. Oceanit's Mind-to-Market approach enabled going from the lab to a manufacturing facility in the Gulf of Mexico in about eighteen months. With this kind of comprehensive, integrated thinking, we can unlock the shackles that hold back disruptive innovation.

CHAPTER 13

Left of Boom and the Intellectual Anarchy Process

Every meaningful discovery opens up new questions that only show us all the wonders that we don't understand.
—Noam Chomsky

"Left of boom" is a term of art used by the US Department of Defense to describe the time before an explosion occurs, or as they say, "before things go kinetic." In innovation, we use the term in an analogous way to describe the flat part of the hockey stick growth curve that all start-ups strive to achieve.

Self-imposed or externally applied pressure to produce immediate impact precludes typical researchers from thinking "left of boom." What we have learned, however, is that sometimes to go fast, you need to go slow. The ability to think about what's "interesting and important," which is left of boom, is where disruptive innovation all begins. Nevertheless, most researchers are more comfortable thinking in terms of what's *required*; that way they can't be accused of being wasteful or inefficient and can defend a near-term return on investment (ROI) for their research investment. While requirements are essential in bringing technology to market, developing disruptive innovation has to start with what's interesting and important. If executed with a disciplined process, it produces a persistent pipeline of disruptive innovation.

You might expect universities to be havens for people devoting themselves to what's interesting and important. My experience suggests the opposite. The first priority of new faculty is securing tenure, after which they'll feel free to pursue what truly interests them. As tenure is heavily weighted in favor of the number of publications and funded research, these new faculty channel their efforts into safe subjects in noncontroversial areas of research in order to hit their numbers. They propose relatively low-risk projects in very narrow fields of study, publishing in thinly read journals reviewed by a small circle of colleagues and associates, creating an academic echo chamber. This has also become a racket for publishers that launch increasingly specialized journals, charging authors for the publication of articles and universities for a subscription. So, the faculty get tenure, the publishers make money, and everyone is happy—except that the level of disruptive innovation is significantly stifled.

Exceptions to Prove the Rule

There are, of course, exceptions to this "play-it-safe" trend. Two current examples I would cite from my own experience are Noam Chomsky and Marc Meyers. Noam Chomsky, known as "the father of modern linguistics," invented generative grammar and developed the connection between language and mathematics, demonstrating an underlying logical consistency to all human language and human intelligence, a radical innovation. Traditional academia initially didn't regard linguistics as a serious subject with adequate rigor. Today, however, its applications are broad—underpinning ideas on how the human mind works or informing NASA scientists on how to communicate with extraterrestrials if we encounter them as humans explore space.

Marc Meyers pioneered the science of biomimicry. He began his career in metallurgy and materials science—the science that brought us steel and aluminum for big industry, factories, and tractors—before pivoting into biomimicry, the study of materials found in nature, including everything from seashells to bones. Marc encountered resistance from people who didn't understand how biology meshed with traditional materials science and thought biomimicry was less than rigorous. He came out with an early text called *Mechanical Behavior of Materials* in 1998, before biomimicry was well established. Today it's an exciting branch of materials science producing all kinds of new materials from artificial bone, tissue, and sensors.

Government Research Priorities Change: Requirements vs. Discovery

Historically, the government undertook the funding of early stage research through a number of organizations. The oldest federal research organization in the United States is the Office of Naval Research (ONR), founded in 1946, which predates the National Science Foundation (NSF), the National Institutes of Health (NIH), and others. ONR and other defense-related research organizations divide research investments into seven different buckets, ranging from Basic Research to Operational System Development.

Research Code	Description
6.1	Basic Research
6.2	Applied Research
6.3	Advanced Technology Development
6.4	Advanced Component Development and Prototypes
6.5	System Development and Demonstration
6.6	RDT&E Management Support
6.7	Operational System Development

Fig. 13.1. Department of Defense research budget activity codes and descriptions.

Over the last several decades, investment has shifted from early stage basic research to late-stage development, a move driven by requirements versus discovery and exploration. The pressure to show relevance in research investments follows a national trend in undervaluing science. This holds for federal research investments as well as national labs and universities.

The focus on the present is causing us to underinvest in our future. You can see it playing out today in the debate over climate change. High-ranking members of Congress dismiss research findings, established evidence, and the opinions of thousands of scientists on the basis that they can pick and choose what science to believe and what to ignore. It's a replay of the battle over tobacco and the connection between smoking and lung disease. Epidemiological studies showed a direct connection between smoking and lung disease, but the tobacco industry funded research that claimed the opposite. The lack of scientific literacy in the country allowed the tobacco industry to present outliers, like a ninety-eight-year old who had smoked every day without problems, as refutations, despite overwhelming evidence to the contrary. If we go further back in history, we find examples of those who manipulated science to show certain types of people were inferior due to the shape of their head or the color of their skin. The lack of investment in basic, fundamental science has a real, human cost.

What's Interesting and Important vs. What's Required

A few months ago, I delivered a Mind-to-Market presentation to a group of about thirty science advisors from the Department of Defense. When I told them that research begins by asking "what's interesting and important," this intelligent and well-educated group surprised me by asking, "How? How do you start your process focusing on what's interesting and important versus what's required?" Requirements are important, as is the ability to generate a return on investment—an organization that ignores requirements quickly becomes irrelevant—but these scientists were curious about how to create technology ahead of requirements, before anyone knew what it was supposed to be or do.

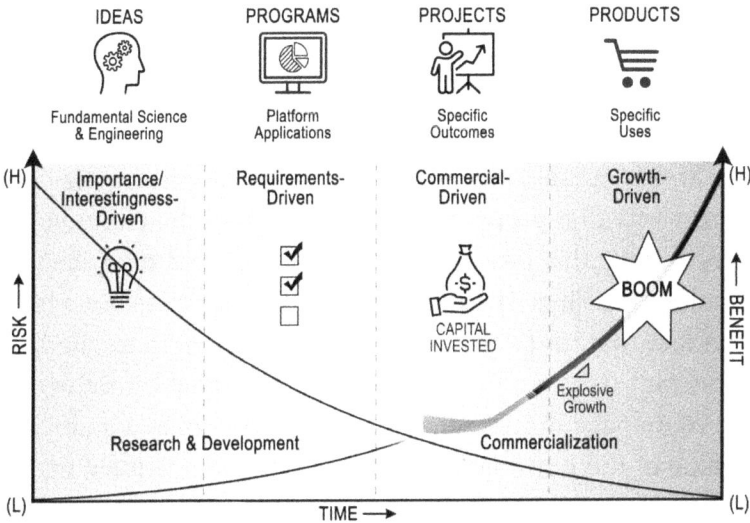

Fig. 13.2. The tortured relationship between research and development and com-mercialization is exacerbated by misuse and misunderstanding of terms. Through years of "efficiency-driven" corporate management and investment strategies, most consider research and development to begin in the "requirements-driven" zone: simply stated, "If there are no requirements, there is no research investment." Going further left into the "importance/interestingness-driven" zone enables seeds of disruptive innovation to be planted. When done with a disciplined process, require-ments will emerge later.

In general, investment capital has no interest in research, where the goal of "commercial-driven" zone is to further reduce risk in technology configuration, utilization, costs, and so on. By the time one enters the "growth-driven" zone, it's all about going to scale and the resultant growth "boom" that all investors are looking for. Nothing about this is trivial. All of these zones have their own flavor of risk. Mislabeling and misunderstanding the zone results in disappointment because of confused expectations and confused execution.

Disruptive innovation starts before requirements, before specific applications are clear or understood. In fact, the first ap-plications anticipated by the researcher are often wrong, and the actual applications become clear only after interaction with the market users. That said, at Oceanit we have a disciplined approach on what specifically to spend time and resources on that is left of requirements and way left of boom. We start by asking ourselves what's interesting and important, then narrow things down to a few areas. Actually leaning into an area helps us generate the requirements that propel the next iteration of the technology.

We repeat this process each year, over and over, building a technology pipeline. The long lead time of basic research is offset by the fact that, once the pipeline is established and filled, there is a constant stream of new technology emerging from the market end. We manage these technologies as a portfolio, with an eye on the value of the portfolio as a whole, and not individual technologies. We don't want to pick winners and losers too early, because a promising technology is subject to forces beyond our control— market timing, global economics, political issues, and "black swan" events. As Bill Gross from IdeaLab showed, market timing is perhaps the biggest issue in successfully bringing a technology to market. A portfolio of technologies is much more resilient to market timing and changes.

Persistent Pipeline

Focusing too much on immediate and guaranteed results skews investment—of money and time—further to the right, closer to the explosive growth part of the curve, where it's more expensive. A technology pipeline allows us to get in earlier and cheaper—left of boom. This is the where the technology pipeline gives birth to disruptive technology.

The challenge for any technology business is to develop a persistent pipeline of new offerings. This is a different problem from a start-up that has a single technology they want to bring to market. Established companies spend a percentage of sales on research and development (R&D), typically around 5–10 percent. Early stage companies may invest much more. I recall talking to the former CEO and founder of Medtronic, Earl Bakken, who mentioned that their R&D budget was 7.7 percent of sales but was closer to 20 percent in an earlier phase of the company.

Research divisions, staffed with highly educated people, are constantly making incremental improvements to their current offerings, but it's difficult for them to consider technology that could disrupt their existing product lines. The problem is that the pace of technological change and global competition is increasing for everyone, so disruption is inevitable. It's not *if* but *when*, and *when* comes around faster and faster.

As already noted, Kodak developed the digital camera in 1975 but didn't act on the innovation since it competed with their chemical film processing business, which owned close to 90 percent of the market. They didn't realize that their business wasn't film per se but capturing images and memories.

This shortsightedness is more the norm than the exception. My experience with large companies is that senior management gets comfortable with what they know and lazy about what they don't know. And they don't want to know. There are exceptions, however.

One interesting example of market-driven innovation is Shell Oil, which was motivated to change by the low price of oil. Shell is an established business, founded in 1907, that came under an existential threat—how to make a living at $50/barrel. The way the industry is organized, the oilfield service providers—Schlumberger, Baker-Hughes, Haliburton—have little or no incentive to change their business practices; they make money no matter how the price of oil fluctuates. Oil companies place bets on particular drill sites; service providers, however, are paid to execute those bets. When the price per barrel is low, that creates huge overhead for the oil companies. The service providers may cut their rates somewhat, but they're not interested in taking this type of risk or making major changes to their business model.

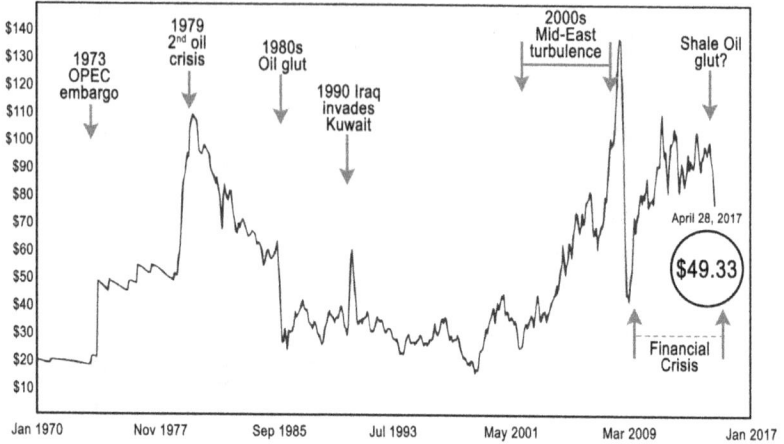

Fig. 13.3. Extreme variations in the price of oil between 1976 and 2017. Very few organizations exemplify market-driven efficiencies more than the energy industry.

At $100/barrel of crude oil, all the constituent parties did well. However, when prices are low, management makes hard decisions, including questioning current technology norms. New methods and materials made available from the advancement of technology in other fields were applied to reduce costs, essentially squeezing the service providers' "book of business." Market pressure is driving the industry to adopt technology that would have been dismissed by prior "status quo" franchise holders. This caused major disruptions through the industry but resulted in much more efficient, performance-driven enterprises. This is an example of what is desperately needed in other industries, including defense and health care, where the government, either deliberately or unintentionally, protects these corporate franchise holders from market forces.

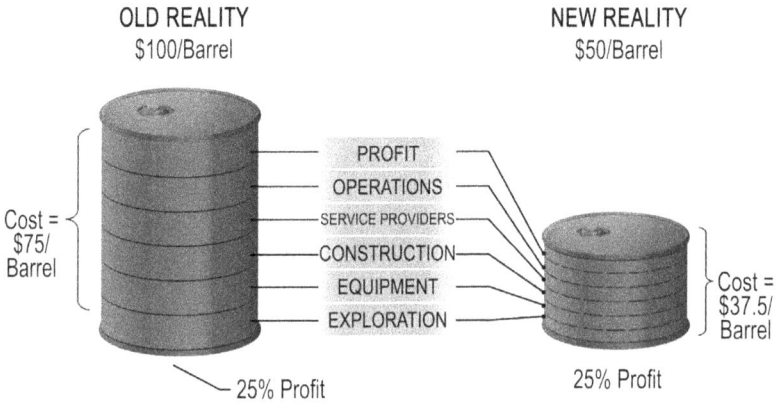

Fig. 13.4. Market-driven change in the energy business. At $50/barrel, change became imperative for survival. Science is enabling change to occur.

In order to turn a profit at $50/barrel, Shell was compelled to invest in technology that would reduce costs. Oceanit was hired to bring disruptive technology to exploration, production, and processing activities to slice off parts of the "book of business" owned by service provider franchise holders.

257

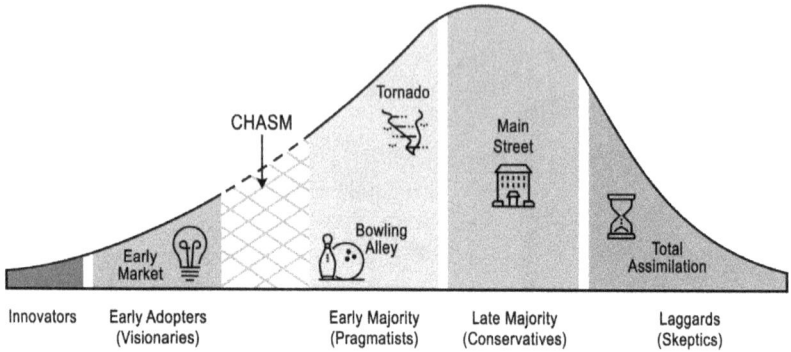

Fig. 13.5. A key challenge in moving technology into the market is referred to as "crossing the chasm" between "early adopters" and the "early majority." It's a major mode of failure for technology adoption, first brought into the business mainstream by Geoffrey Moore in his book, Crossing the Chasm (2002). Note that each market sector has its unique challenges and requires different strategies.

The Intellectual Anarchy Process

> *One of the most exciting and energetic*
> *forms of thought is the question . . .*
> *it illuminates new landscapes and new areas.*
> —John O'Donohue, Irish philosopher

The Intellectual Anarchy process persistently develops disruptive innovation, although it's not always clear where the innovation will lead or what market it's best suited for. However, it's best suited for early market adopters in the market lifecycle because new and disruptive technology typically suffers crossing the chasm, as Geoffrey Moore describes in his 2002 book, *Crossing the Chasm*, shown in figure 13.6.

At a high level, the Intellectual Anarchy process starts with asking a question—something that's interesting and important. This is a broad area but can include just about anything or any industry. Sometimes the best questions are the simplest or most obvious. Graphically, the process is outlined in figure 13.7.

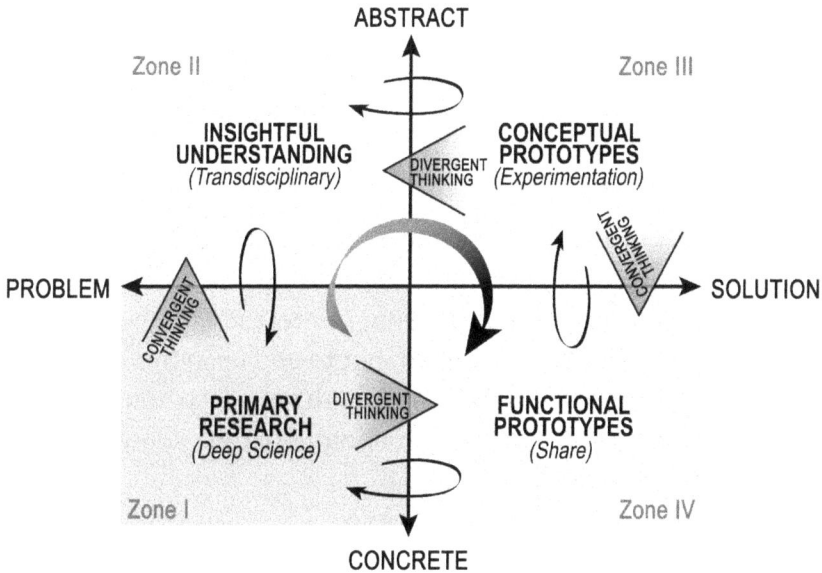

Fig. 13.6. The term "Intellectual Anarchy" was originally developed to acknowledge and explain the messy and chaotic nature of disruptive innovation. Nevertheless, although messy, when managed through a disciplined process it always produces interesting and profound results. Key to practicing Intellectual Anarchy is understanding the nature and expectations of each zone. Even when results are not what was expected, they typically yield a thought-provoking and unexpected insight and point in interesting directions.

We start in Zone 1, the Deep Science zone, with broad or unanswered questions, considering questions that are interesting, important and relevant to big issues or ideas.

Zone I: Deep Science

The challenge of Zone 1 is that researchers are reluctant to ask difficult questions for fear of appearing uninformed in their field. It's easier, and encouraged, for them to stay in their lane and stick with what they know.

To succeed in Zone I, these basic guidelines can help:

- Adopt an explorer's mindset
- Find interesting questions
- Embark on a journey of discovery
- Look at old problems with fresh eyes—use transdisciplinary thinking

The work of the Deep Science zone begins with curiosity. By adopting an explorer's mindset, we relinquish the need for immediate success and can focus on what is interesting. What qualifies as "interesting" is a big topic itself, but to end up in this zone it will usually be driven by geopolitical, health, or social concerns that are part of our annual discussion about what we should do with our time on the planet.

Some questions we've considered:

- How do we distinguish a decoy from an exoatmospheric missile?
- What can we do with nanotechnology?
- Can we control explosions on a nanoscale?
- Can we cure cancer?
- Can we develop materials that don't exist to make space flight more affordable?
- Can we build a prosthetic brain?
- Can we build a space elevator?

The key for Zone 1 is not to limit ourselves to our current expertise. We'll look at cancer, for example, even though we don't have an oncologist on staff. This is important because for a truly disruptive innovation, there is no expertise—it has yet to be discovered.

Zone II: Interdisciplinary
In this zone, we emerge from the Deep Science of Zone I with an interesting question. The next step is to propose an experiment to gain an "Insightful Understanding."

Some guidelines for success in Zone II are to

- Build interdisciplinary teams;
- Think across disciplines/boundaries (transdisciplinary thinking);
- Question authority/expertise;
- Look at extreme examples to frame a problem.

The Zone II experiment tends to deliver a clear, binary answer, a yes or no, on whether the idea is worth pursuing further. It often manifests as an *aha!* experience that grants a curious insight that we may not fully understand but still find interesting.

Actual experimentation is critical to this stage. In the world of engineering academia, there are two tracks to choose from— theoretical and experimental—and most dissertations combine elements of both. Theory can be debated forever; however, a clear, experimental result, particularly an unexpected result, produces the basis for moving forward into Zone III versus falling back to Zone I.

Success in Zone II benefits from an interdisciplinary team because the question being pursued is in new territory. This is particularly relevant in the world of technology, as methods and materials are constantly evolving.

Before performing an experiment in this zone, it's important to review what's already been accomplished by others working on similar projects. There's typically a lot of information available online in journal articles, and so on.

A Prosthetic Brain?

I'm on the Board of Directors of the Rehab Hospital of the Pacific, where I've seen stroke patients with aphasia (the inability to speak), elderly patients with memory loss, and patients with head injuries suffering from cognitive impairment. A sensor that could literally "read your mind" would be an incredible aid to understanding and possibly even assisting these patients in living better lives.

In considering a prosthetic brain, we believed we could read a person's mind by measuring and characterizing the neurons firing in their brain. That capability would depend on having an adequate model of how the mind functions. We found a model in the course of our review that was more complex than what we needed but was a decent starting place for establishing a team to do the work.

We began by developing several "crazy" ideas and explored risk appetite with colleagues in parts of the federal government. In particular, we met with researchers who are trying to give veterans a better life after significant injury. This is a particular issue of modern warfare, where advanced battlefield medicine is keeping soldiers alive with injuries that previously would have been fatal. The good news is these soldiers are living longer than ever before. The bad news is that the road to recovery is longer as well.

Matching risk appetite with a sponsor is not an easy task. Most sponsors don't want risk. Over the course of the past thirty years, I've seen the risk appetite of funding institutions cycle up and down—but mostly down. The careers of the bureaucrats running these organizations are at risk if the project fails, and they prioritize their jobs over their mission to further scientific inquiry. As a result, organizations like National Science Foundation and National Institutes of Health have become more risk averse. They're only interested in funding science if it resembles something they've funded before. This directly impacts national policy issues such as health care. Even though NIH has an operating budget of over $30 billion per year, little has changed in how we fight cancer. That's why Geoff Ling, formerly of DARPA, has been talking about a "DARPA of health care" organization willing to take big risks in health care to break the logjam in affordable and effective health care technology, which is slowly becoming a national crisis.

I grew up with a TV show called *The Six-Million Dollar Man,* about a military test pilot who gains superhuman capabilities when he is "rebuilt" with advanced technology by the military after his plane crashes. With that show as our inspiration, we looked at who in the US government would have an interest in applying our approach to reading the mind.

We discovered that historically there's been a great deal of interest in a brain-computer interface (BCI). Most existing approaches use sensors worn on the head to detect alpha waves or rely on metal electrodes inserted into the soft tissue of the brain. Neither of these approaches is particularly satisfying. The resolution of alpha waves is very low, and the invasiveness of inserted electrodes is unsuitable for most applications. There are other approaches with interesting findings, but no one has been able to develop a really good BCI.

Our interest and approach focused on detecting the magnetic signal given off by neurons firing in the brain. The good news is that the magnetic signal moves easily through the tissue and bone in the skull, so the sensor can be worn externally. The bad news is that the signal is really, really small and is easily overwhelmed by the magnetic field of the Earth or a nearby television set. It's a very hard measurement, but that's where we started.

We found a group within the Department of Defense that was interested in improving the life experience of severely injured soldiers—those with major spinal injuries, missing limbs, or in need of some prosthetic support to have a more normal life. They were willing to take the risk, since the potential payoff would be enormous. Groups like DARPA look at payoff not in venture capital ROI terms but in terms of impact to humans and society.

We proposed an array of hypersensitive magnetic sensors that would rest on top of the head. The first version looked like rollers to curl hair. We also had to create software to drive the hardware and interpret the results. We needed to subtract out the magnetic fields of the environment, so we'd be left with just the magnetic field from neurons firing in the brain. We developed an AI approach to mathematically locate a given signal within a 1 mm cube of the brain, giving us an incredibly high resolution signal—not quite to the level of an individual neuron (we're working on that) but a much higher resolution than possible from an MRI or alpha wave detection. This all had to be done without shielding or cryogenic cooling.

We bought parts on eBay, among other places, to build a pro-totype coil that would provide the sensitivity. This took months of experimentation to convince ourselves that we could make it work.

Zone III: Conceptual Prototypes / Experimentation

The goal of Zone III is to develop a conceptual prototype that demonstrates the validity of our understanding (gained in Zone II) and can serve as the basis for the Design Thinking process in Zone IV. Some guidelines for success in Zone III are to

- Make stuff;
- Try things;
- See how people use;
- Tell a story;
- Promote emotional content—get people excited with the results.

To develop a convincing experiment, we utilized the animal lab at Florida State University that was specifically set up for doing neurological research using rat models. Rat brains share enough key features with humans to provide useful results.

We designed the experiment to place our MIND Electromagnetic Localization Device (MELD) sensors on the external surface of the rat's skull and an electrocorticography (ECoG) sensor inside the rat's brain. Then we stimulated the rat's audio cortex, the region of the brain that shows a correlation between sound and activity. This was a tricky experiment because we needed the rat, and the rat's brain, to function. We went through all the necessary protocols to respect the rat's life. Opening the rat's skull to insert the ECoG required that the rat be partly anesthetized, while being kept comfortable. A team of medical experts that specializes in these types of tests was essential.

The experimental question was: Could we stimulate the audio cortex of the rat's brain and measure that stimulation with our sensors? We "ground-truthed" or verified the correlation with the conventional ECoG sensor. The short answer was, "Yes."

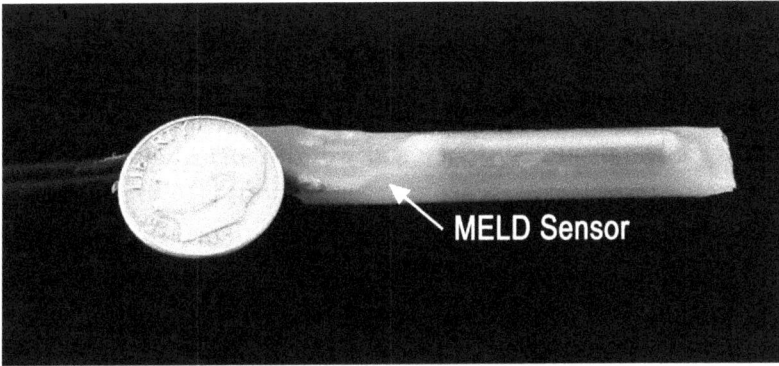

Fig. 13.7. First prototype MIND Electromagnetic Localization Device (MELD) sensor able to track magnetic signals from inside the brain from neurons firing. It operates without traditional shielding or cryogenic cooling required for MRI's and similar technology.

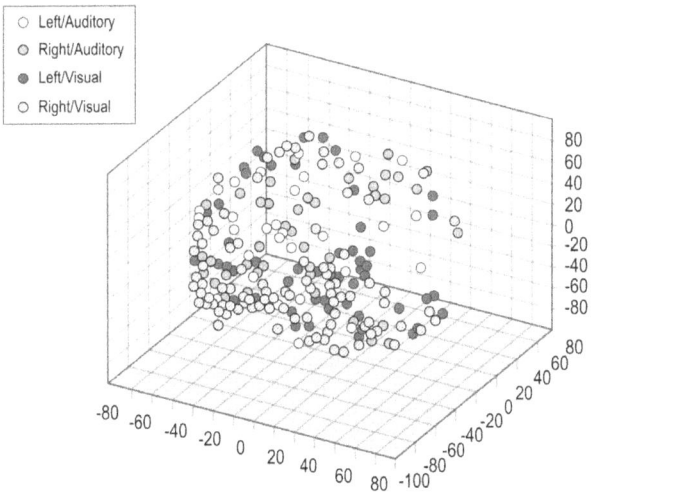

Fig. 13.8. Early results on a test human subject showing signals generated from inside the skull after audio or visual stimulation. When the subject was stimulated, three-dimensional locations from inside the skull were derived from magnetic-field responses, signals that propagated through the brain tissue and bone of the skull to be measured with an array of MELD sensors placed on the head, like a cap. This was not a carefully controlled experiment, but more of a "ready-fire-aim" test, but the results were surprising and interesting.

MELD sensors developed as a sensor for the "Prosthetic Brain" program demonstrated the ability to measure extremely fine magnetic signals without cryogenic cooling or electromagnetic shielding, which is required for medical MRI measurements. Whereas the electric signal from internal brain neurons is unmeasurable on the outer skull, the magnetic signal from the neurons simply passes through the brain, skull, and other tissue. The initial focus of MELD was focused on tracking groups of neurons located within a 3-mm volume inside a skull without penetrating the skull. Preliminary results of the "rat test" show that the MELD sensor was able to measure the audio stimulation in the brain of a rat, correlated with ECoG sensors simultaneously inserted into the rat's audio cortex. Although more work is required, this profound result showed that the magnetic signal of neurons firing from deep inside the skull can be measured and analyzed in an open environment, inexpensively and without penetrating the skull. Somewhere in the future, a new brain computer interface will be available, affordable, and as simple to use as putting on a baseball cap.

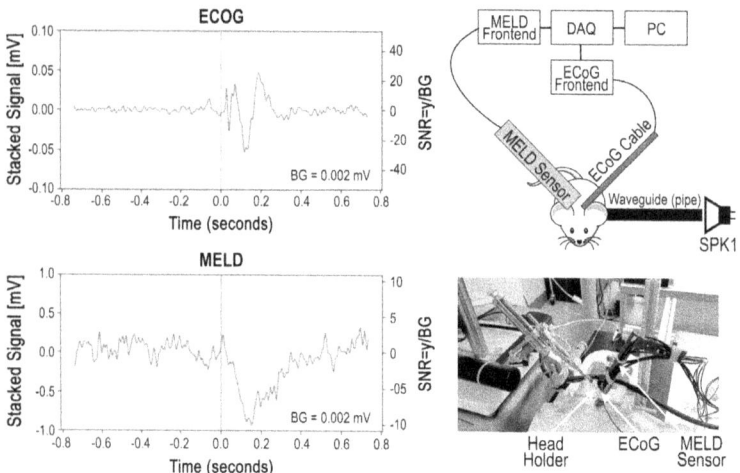

Fig. 13.9. Early results from rat testing, mapping response in the audio cortex with the MELD sensor, simultaneously with an implanted ECOG sensor, in response to an audio stimulation. (Oceanit)

266

As you can see from the figures and results, this stage of research requires enormous focus from highly educated researchers to provide a fundamental yes/no answer to an interesting and important question.

Zone IV: Functional Prototypes / Sharing

This is where we can build a product that will work in the chosen environment for the specific application that the market wants.

The principles for success in Zone IV are to

- Practice empathy (Design Thinking);
- Push innovation to market (Mind-to-Market).

Once we knew that the science worked, we began the journey to find the best application to focus on, discussing various applications with end users and corporate codevelopment partners. The journey to the marketplace is always long, with many twists and turns.

The MELD sensor has the potential to create a high-resolution brain-computer interface with applications in wildly different fields. Applications we're looking at in medicine include controlling prosthetic limbs, counteracting the effects of Parkinson's disease, and gaining a better understanding of Alzheimer's. For computer gaming, especially virtual reality games, the BCI could be used as a controller, creating a seamless, immersive user experience. And for computer-assisted training, the BCI could enable a hands-free interface for manipulating and exploring complex objects in 3-D.

The extremely sensitive MELD sensor also has applications outside of a BCI, as a tool for underground oil and gas exploration, to locate items inside the body during arthroscopic surgery, or as a navigational device to traverse the planet without the use of GPS satellites. Each of these approaches has us exploring "market pull" with candidate users.

PRELIMINARY HUMAN TESTING

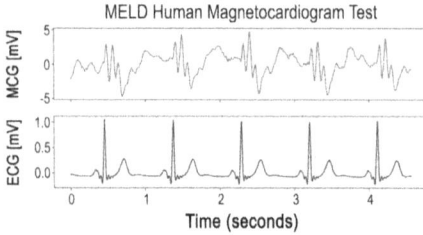

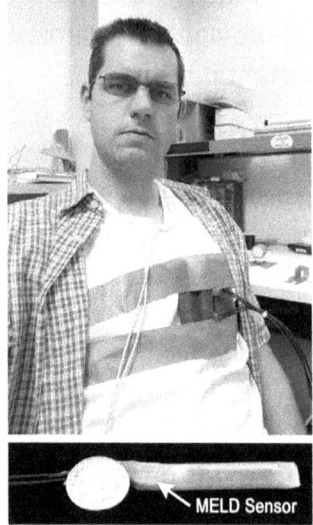

Preliminary Tests & Results

- Oceanit has successfully demonstrated that MELD is capable of measuring a live human heartbeat
- Unshielded & uncooled operation
- Compact wearable form factor
- Approaching millimeter scale resolution
- Undergoing additional real-world human testing

Prototype Sensors

Fig. 13.10. Preliminary human testing results of MELD sensor. Results show the magnetic signal from the sensor taped over the subject's chest, illustrating the potential as a noncontact EKG type of sensor.

These four zones cover the basic process. The way one steps through the zones depends on many other variables, including what's discovered or determined in each zone. This process illustrates the basic elements of Intellectual Anarchy and the benefit of asking an interesting and important question.

//////////

Implications

Somewhere in the future, people with cognitive challenges will be able to rely on a prosthetic device to give them more functionality—what we refer to as the "Prosthetic Brain." This could be manifested as a device to help degenerative brain disorders like Alzheimer's or dementia; or if an objective measure of pain can be developed to navigate between physical pain and mental pain, perhaps as a better tool for pain management.

In the near future, we see a tool that could extend productivity as a brain-machine interface, controlling manufacturing, processing plants, aircraft, cars, or other machines. We can also see a brain-computer interface that will expand gaming, virtual reality, or otherwise optimize immersive simulations with constant feedback to connect reality, perceived reality, and fantasy.

The extremely sensitive MELD sensor also has applications outside of a BCI as a tool for understanding planetary conditions, perhaps reducing unexpected volcanic eruptions, oil and gas exploration optimization, or even as a medical tool to more efficiently execute arthroscopic s surgery—navigating the 3-D space of a body—all based on historically unmeasurably tiny and otherwise unknown magnetic field signals, now possible with MELD. Eventually we see this type of technology enabling terrestrial or space navigation without GPS satellites. Each of these approaches already has us exploring "market pull" with candidate users.

///////////

Ask a good question: this is the key to disruptive innovation. A good question starts left of boom and left of requirements. Most organizations don't have the discipline or patience to indulge this practice. It's easy to see why: just look at popular press—if you are a start-up and don't become a Unicorn ($1 billion valuation) in a few years, then you must be a failure. Nobody wants to be branded as a failure. Venture investors want immediate gratification; there are very few investors who bring disciplined patience—there is only one Warren Buffett. Although Buffett is not known for technology investing, his mentor Benjamin Graham, the "father of value investing," was a big advocate of investing with discipline and patience—also necessary for investing in "value creation."

Public companies are punished if they don't show consistent value growth and returns. If they make a misstep or miss a projection, they must be a loser. People get angry because they have grown to expect near-instant gratification, since that's what the "smart guys" get. Government is not far behind this, supported by politicians who struggle to think long term, since everybody wants immediate results. This mentality undermines what it takes to produce disruptive innovation.

What makes this harder still is the hostility to science finding its way into public discourse—from climate change to vaccines. The issue is exaggerated by social media, since just about anybody can say just about anything on social media, and the considered judgments of scientists and engineers are no more trusted than the opinions of political pundits.

As a result of this short-term mentality, innovation now typically starts with "requirements." Give me the requirement, and I'll give you a product. This approach has its place, particularly when it comes to paying the bills. That's why at Oceanit we think of our work as projects distributed in a portfolio of risk, in three buckets: interesting, challenging, and disruptive. The first two buckets have low risk and pay the bills. The third bucket, where the risk is high, can provide long-term value. We try to pay the bills with interesting and challenging projects, mostly requirements driven, but this does not lead to true disruptive innovation. Managing this portfolio discipline is difficult, but it's this discipline that enables us to pursue disruptive innovation.

Left of boom starts left of the hockey-stick growth we all expect, left of instant gratification. It goes to a place few are comfortable with: the big elementary questions. What should we do? What's important and impactful to make the world a better place? These questions are as much philosophical as technical. We've exercised this process for the last several decades, typically in quarter four of a calendar year, as we prepare for the next year. We've adopted a Darwinian business philosophy: the world has changed, so how do we (the business) evolve? What should we think about? What projects should we keep, kill, or build? By creating space to ask important questions, we pretty much always find interesting benefits.

This is where "left of boom" starts and the "Intellectual Anarchy" process begins. The Intellectual Anarchy process provides a disciplined framework to journey through an inherently chaotic process of exploration and discovery. It seems somewhat counterintuitive, since we need to explore the unknown without having a clear destination.

However, I have consistently found that asking an interesting and important question always produces new insights with potentially disruptive implications. What we cannot predict is just what these implications are for humans and society—only that they will be impactful and produce new opportunities. The result is a pipeline of technology that is both disruptive and impactful.

This journey is not really anarchic, but it may feel somewhat anarchic because it attempts to deliberately find connections between the academic disciplines we have been educated to respect but that often become limited by their artificially imposed boundaries. It's only when we get comfortable exploring the edges of science and knowledge that we can unlock the mysteries of the universe. We may disrupt everything around us as a result, but this approach can accelerate finding solutions to some of the world's most perplexing problems.

PART V
Disrupt

If you don't know what you can't do,
sometimes you can do the impossible.

CONCLUSION

*Invention, it must be humbly admitted, does not consist
in creating out of void, but out of chaos.*
—Mary Shelley, author of *Frankenstein*

Intellectual Anarchy is a process that persistently produces disruptive innovation. "Disruption" is a term loosely used and widely celebrated, but make no mistake, not everyone wants to disrupt, and no one wants to be disrupted. Companies and individuals that control the status quo, the franchise holders, have the most to lose. Nevertheless, disruption is as fundamental to business as Darwin's theories of natural selection, evolution, and survival of the fittest are to nature. For business, it's not a question of if but *when* disruption will occur; it's an inevitability. Today, as the kinetics of business occur faster and faster, disruption needs to be anticipated and planned for.

Broadband Internet disrupted copper-wire telephone service, forcing phone companies that had heavily invested in copper wire to change—a painful process; workers were laid off, and investment into training and new technology was costly. Today, major disruption is occurring with electric companies who find themselves of two minds, proclaiming their allegiance to clean, renewable energy while desperately clinging to old, profitable standards. Before that, electric light disrupted gas light in the home. Resistance to disruptive technology is nothing new.

Even militaries suffer from an aversion to disruption. Once a system is in place, in use, and part of military training, the cost of change is difficult to swallow. If not for the threat of adversarial military superiority, little would change; we'd still be lobbing cannonballs at each other. Militaries that attach too great an importance to prior investments or commitments can be reluctant to

275

evolve, choosing instead to defend their adopted mindset. For those that refuse to change, it typically ends badly. In *Lifting the Fog of War*, retired admiral William Owens outlines how, in the Franco-Prussian War of 1870, the Prussian military faced a superior French force with greater numbers, more combat experience, and better weapons. Whereas the rest of Europe was investing in railroads, the French clung to their past investment in canals and water transport infrastructure and were reluctant to adopt rail into their operations. This worked to the advantage of the outgunned Prussians, who exploited the civilian railroad, handing them a clear victory over the French.

Some organizations weather this change better than others by embracing self-disruption. Apple is probably the best example of this. Steve Jobs disrupted the successful Apple II computer with the introduction of the Macintosh, a move that seems obvious in hindsight but was wildly controversial at the time. Jobs was fired from his own company, in part due to the Mac's failure to live up to his predictions. Twenty years later he was rehired and turned Apple, and the world, on its head with the introduction of the iPhone, and then disrupted the Mac in turn with the introduction of the iPad (declaring that the world had entered "the post-PC era").

IBM disrupted itself with the introduction of the IBM PC, a product that succeeded far beyond the company's expectations or desires, and then again when they pivoted from hardware into services. Kodak, on the other hand, invented the first digital camera, but chose not to self-disrupt, to protect their chemical film processing franchise. One must wonder, if the CEO of Kodak could have seen the future, or if he had read this book, would he have done anything different? Would he have risked destroying the company's profitable, existing business, with close to 90 percent of the global market, to pursue digital? We'll never know.

Disruption is hard and should be considered a contact sport. It's not for the faint of heart. There are winners—those who control the disruptive technology, and everyone who benefits from improved results and lower costs—and there are losers—the current franchise holders, who fight vehemently to protect the status quo. Nevertheless, disruption is an inevitable part of the creative destruction of the business cycle, the accumulation and annihilation of wealth. We must disrupt, therefore, for two principal reasons: (1) it's essential to our survival, and (2) we deserve better.

It is the time to dare and endure.
—Winston Churchill

Winston Churchill was a disruptive choice to lead the British in World War II. London had been pulverized by the Blitz, and England was not expected to survive. Many Britons believed that surrender was the only option, but, backed into a corner, they made a disruptive choice, turning to a nontraditional leader who declared, "We shall never surrender." It was a choice driven by desperation, not logic. However, with help from Roosevelt and the United States, Churchill was able to restore British resolve and took the fight to the Germans, against overwhelming odds, turning certain defeat into victory. Would a traditional leader have done the same thing? It's hard to say. But I would wager that Churchill's audacious statements, as implausible as they were, were key to turning the tide.

We must disrupt to survive, because that's the only path forward. When we face an existential threat, we must be willing to risk it all when we choose a path forward. In the 1998 movie *Armageddon,* an asteroid is on course to hit Earth and destroy all life. A desperate world trains and launches a team of unlikely heroes, a group of oil-field roughnecks led by Bruce Willis, on a mission to land on the asteroid and place nuclear warheads to break it apart and save the planet. It's a desperate plan, with a low probability of success, but the calculus is clear: the cost of failure (a few billion dollars and the lives of a handful of roughnecks and astronauts) is less than the cost of doing nothing (the end of the world).

So, too, with disruption. The cost of failure (a "wasted" R&D budget and missed opportunities) is less than the cost of doing nothing (getting blindsided as a nimbler competitor eats your market).

Not every disruptive technology involves the fate of the world, as the following chart illustrates, but for those willing to embrace the risk, they deliver the biggest payoff. We've developed a framework, Intellectual Anarchy, for thinking about disruptive innovation and where to look for categories worthy of the challenges and difficulties of bringing disruption to market. We have found this to be one of our best targeting tools.

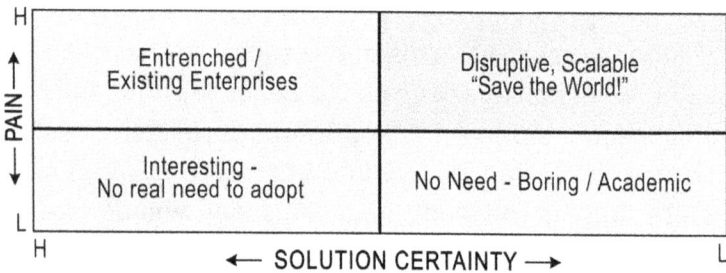

H	Entrenched / Existing Enterprises	Disruptive, Scalable "Save the World!"
PAIN	Interesting - No real need to adopt	No Need - Boring / Academic
L		

H ← SOLUTION CERTAINTY → L

Fig. C.1. When considering what constitutes disruptive innovation, framing where to look is helpful. We have used this tool for many years. The realization that one needs to investigate areas with "high pain and low certainty" is counterintuitive for most. However, it's where resources meet need. Once we realize that almost everything that matters is difficult, sometimes impossible, we take that for granted.

The second broad categorical reason to disrupt is that we, as a community, deserve better. We deserve safer, cheaper, more effective products and services. We deserve a clean environment where climate change is not a threat. We deserve to have better health care that is both effective and affordable. All these things are doable, but they require that we take risks and try new things, collaborating across multiple disciplines.

Disruption isn't limited to science and technology; it can include everything from business models to finance. Uber, for example, leverages a disruptive business model; their technology stack is nothing special. In the process they've created a more convenient, more efficient, more pleasant alternative to taxicabs.

In places like New York City, the taxi industry is a stagnant monopoly, constrained by the need for a taxi medallion, which are in limited supply and only available via auction for as much as a million dollars. How does a potential new entrant justify the cost of a medallion when taxi fare ranges from $10 to $100? Uber turned this model on its head by requiring its drivers to work as contractors who supply their own vehicles, radically lowering the capital costs and enabling rapid expansion.

One of the things I like most about Uber is that it works anywhere in the world. I've used the service throughout the United States but also in South America, Europe, and the Middle East, with a uniform experience despite the disparate languages and currencies of all those areas. A few years ago, on a business trip to Rio de Janeiro, I caught a taxi from the airport to my meeting. The driver was nice but didn't speak English. We were able to communicate well enough in Spanish—not his first language or mine—that I reached my destination, but I'll never forget the smell of the cab itself. It smelled like a crime had been committed in the back seat and then hosed out before the next customer.

I took an Uber to my next meeting, and I was picked up in a clean, relatively new car by a driver eager to practice his English. We talked about the post-Olympic economy of Rio and economic outlook for Rio's youth. It was a completely different experience. I've become a fan of Uber wherever I travel, with similar experiences wherever I go. Uber has disrupted the taxi industry. That's bad news for NYC medallion holders but good news for everyone who benefits from a cleaner, safer, more convenient ride.

We're gambling on our vision, and we would rather do that than make "me too" products. Let some other companies do that. For us, it's always the next dream.
—Steve Jobs

While this book is intended to serve as a roadmap for disruption, not every organization will be capable of following the same route. Some lack the capacity. Some are ruled by fear or simply lack the will. Some are deluding themselves into thinking they're pursuing disruption when they're not. Others are too invested in their current success to take a chance. All are at risk of being disrupted by someone else.

To deliver disruptive innovation you need three things:

1. People
2. Culture and Environment
3. Organization

People | Culture & Environment | Organization

Fig. C.2. For practicing Intellectual Anarchy successfully, there are three essentials. First, one needs the right people. Second, establishing the right culture and environment is critical. Third, an organization must empower these highly educated and talented individuals to apply their skills. Instead of the conventional "command and control" organization, a more horizontally structured organization with limited micromanagement is required. It's a difficult balance—metaphorically, the organization needs to enable great artists to play jazz versus play sheet music.

People

An educated workforce is table stakes for technology innovation. Disruptive innovation requires bright, highly educated people. We tend to get our best results with young people—not so much young in age as young at heart—people who are naturally curious, who have not given up their youthful outlook of the world.

I fly for work more than I would like to, perhaps several times per month. Though I sometimes tire of travel, I've never lost my excitement for airplanes as a technology and the concept of heavier-than-air flight. Sometimes I like to just watch the separation of air around the wing—it's applied hydrodynamics but with a compressible fluid. For me, airplanes are cool and always have been.

We recently hired a young researcher who was finishing up his PhD in physics. He was a bit scruffy and awkward at his initial interview—not your typical candidate. Most people would pass on him, on the basis of his appearance and his resume, which didn't present well. But I learned that his parents were former Soviet scientists who immigrated to New York when he was a child, and that he built his own transportation, his Kawasaki KZ1000 motorcycle, from spare parts—facts not on his resume. I could see that he was a very intelligent and resourceful young person, motivated to do something with his life. Those who have worked with him are amazed. He's fast and focused, yet open-minded and fearless—a true Techno Warrior in the making.

Culture and Environment

There's an innovation strategy that supports the idea that beautiful buildings and efficient spaces create the environment for innovation. It's an aspirational strategy in the style of *Field of Dreams*—if you build it, they will come. Significant capital is deployed to create tastefully appointed tech hubs and incubators, when it's actually the culture and people that will inhabit those spaces that matter. Culture, not environment, is the driver of innovation.

To be sure, environment reinforces culture. So, perhaps better said, culture drives innovation more than tenant finishes. When environment reinforces culture, culture becomes part of the operation.

Creating the right environment doesn't have to be expensive. In a recent meeting with the dean of the College of Engineering at a large, public university, I was asked about tactics for getting people from different disciplines to work together more, to reinforce a transdisciplinary culture. One simple suggestion is: great coffee! Provide a shared lounge with amazing coffee that doesn't belong to any specific department. This small change would create an environment for people to have conversations with people they would otherwise never run into, since they have separate elevators, labs, corridors, and so on.

I recently had the privilege of visiting the bustling capitals of the Middle East on business, but I took a few days on my own to visit some less developed regions to meet people in small towns and villages. In Oman, for instance, I met a young man who was working with tourists but aspired to save money and get married, much like young men I know in the United States. Education was an important part of his plan, and he had taken the time to attend college. He spoke English well and understood the challenges of growing into the future while remaining in touch with his culture and his family values. In some ways, he could have been from the countryside in Colorado, Kentucky, or Tennessee.

Dubai, in contrast, is a massive development that springs from the desert like Las Vegas on steroids. It's a beautifully designed city with amazing architecture—clean, efficient, and expensive— with the elegant Burj Khalifa tower, a building over half a mile tall, as its centerpiece. Here the similarity with free-wheeling Las Vegas ends. In addition to a prohibition on gambling and restrictions on alcohol, Dubai is a sovereign, absolute monarchy, ruled by Sheikh Mohammed bin Rashid Al Maktoum. Dubai is one of the seven federated states of the United Arab Emirates (UAE), along with Sharjah and Abu Dhabi, which serves as the capital.

The Emirates are intent on diversifying their economic base and modernizing their infrastructure. They speak of pursuing disruptive innovation in pursuit of these aims, but their efforts are hampered by their culture.

> *If we knew what we were doing*
> *it wouldn't be called research.*
> —Albert Einstein

I met with a group at the American University at Sharjah that wants to develop a more robust research community to produce disruptive innovation. The campus is spectacular, like a Stanford

University in the desert. When I asked about fundamental research, however, or high-risk research where the answer is uncertain, I discovered that they weren't really interested; they were more focused on reducing risk to minimize criticism, since the university is supported by an absolute ruler.

This isn't real research. You can't know the answer before you ask the question, and the desire to know in advance eliminates the possibility of true exploration or discovery. While there is enormous value to teaching and mastering the fundamentals of science, math, and engineering that have evolved over the centuries, reproducing the classic experiments, and so on, it's not the kind of high-risk research essential to disruptive innovation. It's difficult to take risk in such an environment.

I had similar meetings in Dubai about their "Disrupt Dubai" initiative, geared toward the adoption of innovative technology. The major emphasis was on bringing innovation to social services and infrastructure—water and power, police, transportation, health care, and so on. While there are advantages to authoritarianism in terms of expedience, it's ultimately detrimental to true disruption.

In the United States, bringing innovation to city services is a nightmare of political compromise. Something as simple as eliminating carbon paper in favor of digital records could be derailed if the CEO of the carbon paper company was friends with the mayor's wife, or some such thing. With the authority of an absolute ruler in Dubai, however, officials who run social services were put on notice and required to adopt innovation.

That said, switching from carbon paper to digital records might be genuine innovation and would certainly disrupt the carbon paper industry, but I wouldn't consider it "creating disruptive technology." That requires true risk and higher stakes. I applaud Sheikh Mohammed bin Rashid Al Maktoum's efforts to make city services more efficient and to create start-up jobs in the region, but there's no way to create true disruptive innovation—which requires taking risk—when the cost of failure is too high, making it impossible to learn from those failures.

True disruption in the Middle East would be something like creating a renewable energy source that replaces fossil fuels and saves the planet, but this would challenge the region's dominant and highly dependent position in petroleum and gas.

China is also emerging as an economic powerhouse. Chinese lead the world in refereed journal publications and are investing significant capital in research. I'm often asked if they are a threat to the United States in the area of science and technology. The short answer is "yes," for multiple reasons, but with caveats. They suffer from the same limitations of the UAE in undertaking disruptive research due to their authoritarian regime. For example, China is rapidly building machine learning–based artificial intelligence (ML–based AI) capabilities, an approach that benefits from large data sets. Their police-state control gives them the opportunity to collect massive amounts of data from an enormous population. But ML–based AI suffers on edge cases and outliers. No dataset is large enough to resolve all ambiguities. And China is unlikely to develop the sort of universal AI that can cope with ambiguity in humanlike fashion.

At Oceanit, we're working with Noam Chomsky on a radical new approach to artificial intelligence. Dr. Chomsky is a great thinker and a professor emeritus of MIT, but in 1973 his outspoken criticism of President Richard Nixon landed him on Nixon's list of subversives. In China, the same behavior would have led to him to being imprisoned or executed.

If you systematically eliminate people who think differently, you condemn yourself to copying research, publishing incremental changes, and making minor innovations. Enrico Fermi, known as the "architect of the atomic age" and a key figure on the Manhattan Project, emigrated to the United States during World War II to escape the Italian racial laws that threatened his Jewish wife. The big changes that will truly disrupt the world come from different views. If we're all forced to work, act, and behave the same way, then there's no space to fail or take risks—and no opportunity to innovate and disrupt.

Organization

Most organizations are designed to be managed, and under-standably so: a poorly managed business does not survive long. They excel at putting round pegs in round holes—Management 101. That works for traditional businesses, but businesses built to inno-vate are inherently difficult to manage. Keep in mind that great busi-ness thinkers like Alfred Sloan (*My Years at General Motors*), whose management practices are still being taught in MBA programs to-day, developed those practices to address emerging industrial op-portunities. They were innovations at the time, but the world has changed, and companies are not well equipped to deal with fast-paced, persistent innovation. True innovation is messy and gritty. Some things work, some things fail; both result in learning. Pursu-ing high-risk projects without a clear payoff in uncertain markets is rarely supported in government or business. However, by running toward chaos rather than away from it, we have created a discipline that enables us to put round pegs in square holes.

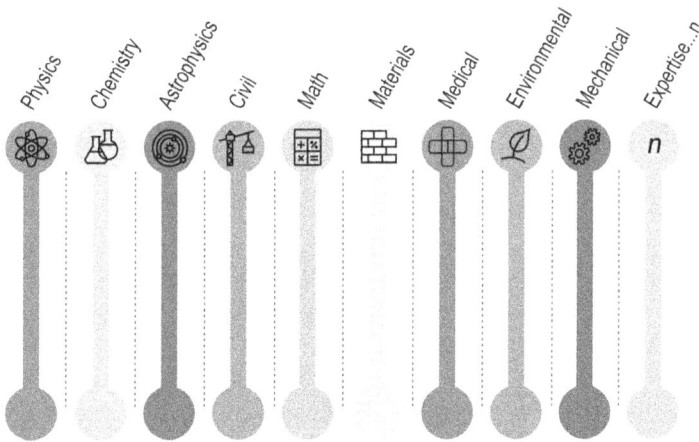

Fig. C.3. Typical research and development organizations benefit from a highly edu-cated and skilled workforce, trained in specific expertise. However, they are typically mandated to stay in their lane.

Unlike traditional organizations, we rely more on people's ability to think than on the domain knowledge they learned in school. We assume they can look up anything that's been published, so regurgitating facts is of little real value. We ask people to set aside what they think they know about their field so they can lay fresh eyes on old problems. We ask them to wade in to new fields and disciplines. Some people are put off by this, but those who learn to operate in this transdisciplinary mode can get very good at it. They get comfortable with the skill of thinking and the application of fundamentals.

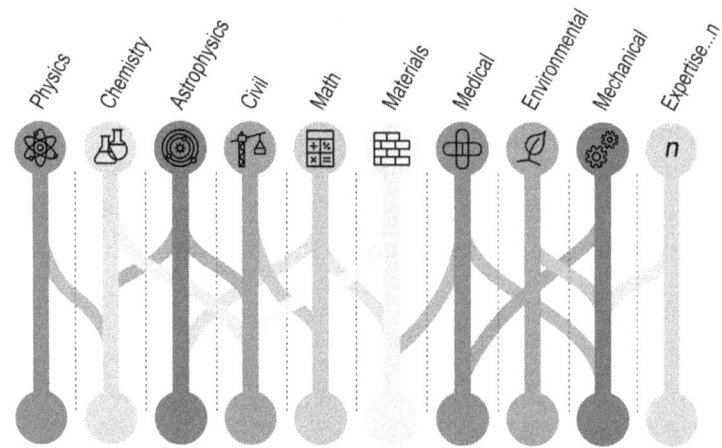

Fig. C.4. Transdisciplinary organizations are built to innovate but difficult to manage. However, in a connected and fast-moving world, where one can access knowledge and facts, review journal papers, or access sophisticated reports in seconds, the key skill is the ability to think. A transdisciplinary organization benefits from T-shaped experts who bring deep domain knowledge as well as broad interests. This is ideal for disruptive innovation, which mixes the skills of a disciplined investigator with the "fresh eyes" of a skilled thinker from outside the field. These fresh eyes will take a look at what may have been an intractable problem, oftentimes yielding new ideas, insights, and approaches.

Operations need to have clear goals and guidelines. At Oceanit, we have a disciplined method to measure performance across several categories, and we're constantly trying to improve on strategy, tactics, and metrics, which are always up for scrutiny and discussion. This Balanced Scorecard approach is developed for each of the business subgroups each year and reviewed quarterly throughout the year.

However, we have also adopted a Darwinian view of our business that asks, "Since the environment is changing, how should we change with it?" In the fourth quarter of each year, we ask ourselves, "What should we do with our time on the planet?" This conversation goes for the entire quarter, wrapping up at the end of the year, when we decide what to keep, what to kill, and what to add to our project portfolio based on our new current understanding of the world, geopolitics, science, the economy, and so on. This has us constantly adjusting our direction to address the market—which we take to mean the planet, not a particular region.

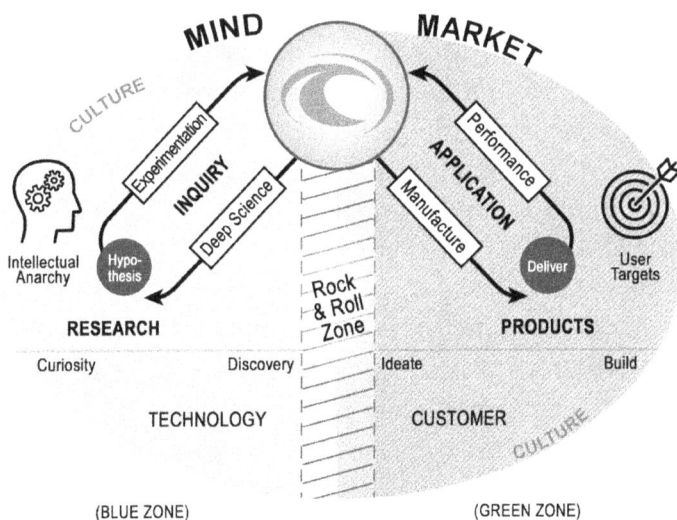

Fig. C.5. The overall Mind-to-Market business model connects the cultures of the Blue Zone (science and discovery) and the Green Zone (application and customer satisfaction). Ultimately, this is how the disruptive technology in the Blue Zone becomes a product in the Green Zone.

We also understand that to bring technology to market, we must use Design Thinking to manage the cultural divide between deep science and discovery on the one hand and human impact and product delivery on the other. We have mapped out the complete ecosystem into three zones. On the left, what we refer to as the Blue Zone, is where deep science, exploration, and discovery occur. On the right, the Green Zone, is where technology delivery occurs, impacting humans and society. In the middle is the Rock & Roll Zone that connects these two cultures.

Typically, these two cultures don't coexist in the same organization; it's one side or the other. Either they're all Blue, like universities or national labs, or all Green, like the Stanford d.school or companies like IDEO. Blue Zone organizations explore science and discover new things, but this new knowledge rarely sees the light of day. It may get published in a journal and read by people in the same field yet have its potential unrealized for decades. Green Zone organizations, on the other hand, excel at human-centered design and refining existing products, improving everything from shopping carts to toothbrushes to can openers. While their designs are often inspiring and artful, they generally don't integrate profound new insights of physics.

We find that utilizing Design Thinking as a common language allows these disparate cultures to collaborate. Often the folks from the Blue Zone, when they've discovered an amazing scientific insight, believe the hard work is done—those Green Zone guys have it easy. They don't realize that commercialization is just as difficult. The nonlinear mapping between deep science and human-centered impact of technology and products that takes place in the Rock & Roll Zone is full of twists and turns, full of surprises and discoveries that make it challenging to actually deliver usable technology. Researchers in the Blue Zone may imagine they know how the market will adopt their brilliant discovery or insight, but they are almost always wrong. Only after spending time with the intended users or candidate users do we discover what they actually need. That's why the Rock & Roll Zone is so difficult to navigate. Particularly with disruptive technology, what people think they want may not be what they need. We call those that work in that zone "technology Sherpas" because it's an activity fraught with danger, challenges, and extreme conditions.

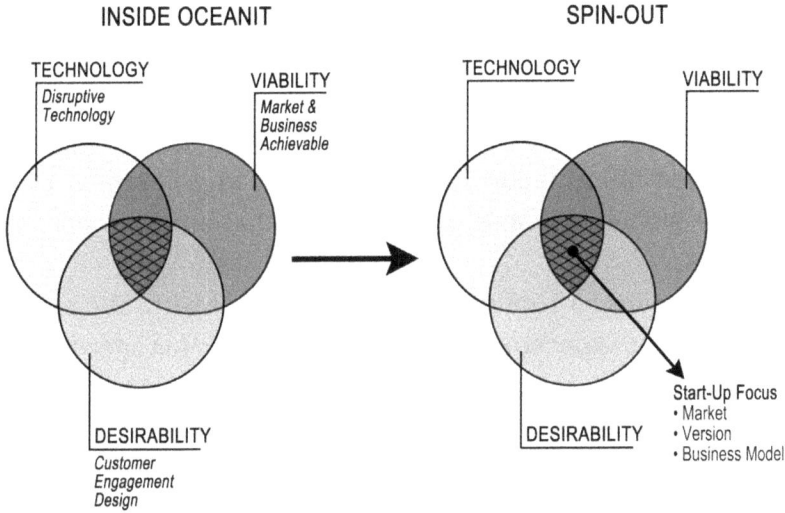

Fig. C.6. Market adoption and the opportunity for disruptive innovation lies at the intersection of technology, viability, and desirability.

Ultimately, we are trying to discover the convergence between technology, viability, and desirability through conversations with end users. We try to empathize with what customers or partners are trying to achieve, versus trying to sell them something they cannot use. The outcome is never obvious.

I think there is a world market for maybe five computers.
—Thomas Watson, President of IBM, in 1943

Recently I was asked about quantum computing—what's the return on investment? What are the job opportunities? What is the economic impact likely to be? It's far too early to tell, but consider the impact of the transistor. William Shockley, John Bardeen, and

Walter Brattain developed the solid-state transistor at Bell Labs in 1946 as a more robust alternative to the vacuum tube. (They received the Nobel Prize in Physics in 1956 for their efforts.) At the time, it wasn't clear what the transistor would mean to the world, but it ushered in the era of modern electronics—everything from portable radios to personal computers. Later, Shockley moved back to Palo Alto, because that's where his family was; Silicon Valley, as we know it, didn't exist. Shockley became a professor at Stanford, and that's when the seeds of Silicon Valley were planted.

Quantum information science has the same potential to rewrite the global economic order in ways that are unfathomable to us today.

In the future, disruption won't take an act of faith. The hardest thing we're up against is that when people ask for specifics, we can't tell them. A century and a half ago, if someone had asked Alexander Graham Bell what we would be able to do with the telephone, he couldn't have begun to predict the ways it would affect the world. He couldn't have imagined that you'd be able to call up relatives on the other side of the world, or order groceries delivered to your door, or find out what movies were playing that night.

What use could this company make of an electric toy?
—William Orton, president of Western Union
Telegraph, declining to purchase Alexander
Graham Bell's telephone technology

Before the first public telephone was inaugurated in 1844, people had discussed the possibility of putting a telegraph on every desk in America to improve productivity, but it wouldn't have worked. The telegraph required its operators to spend forty hours or more learning to transmit and receive the arcane sequences of dots and dashes that make up Morse code. Bell's telephone obviated the need for special training. People already knew how to use it—you just talked.

Methods and materials change. What was once impossible may become possible—today or in the future. The implications of this simple statement are profound. Many of the problems perplexing the world today seem unsolvable but may in fact be solvable, or at least addressable. Geography need not be the limiting factor it once was, as long as we have educated people with access to electricity, transportation, and connectivity, the essential building blocks of the future. Some of the biggest challenges we face today fall into these buckets:

- *Climate*. We must decide to engineer a climate solution based on solid science and causal understanding. It's not enough to just take data and show correlations; we must understand causality to engineer a planetary solution.
- *Social inequality and poverty*. The world can certainly do better. To address mass migration, income inequality, and social injustice, we need to innovate economic opportunity.
- *Brain drain*. This is a problem all over the world. Most people want to live and work in the place where they were born. We must disrupt to stop the brain drain in communities that seem to offer little opportunity to build a future for young people born and raised there.
- *Health care*. The system we have today was built a century ago. It rewards the franchise holders and transfers the expense and cost to the community. An idea that's been bouncing around Washington, D.C., is a health care version of DARPA—HARPA (Healthcare Advanced Research Project Agency)—to encourage and support high-risk, high-payoff science and technology development in medicine and health care.

Given that disruption is not only inevitable but necessary for our survival and to build a better future, this is the question you must ask yourself: Is my organization the one that gets disrupted? Or the one that disrupts? And if you choose to disrupt, I hope this book has helped to point you in the right direction.

ACKNOWLEDGMENTS

Nothing occurs in a vacuum; it takes an ecosystem. The concepts and principals discussed in this book are informed by many amazing, bright and talented people at Oceanit. They have been willing subjects in the living experiment to find the edges of science, or summon the courage to examine or unravel what may seem like unsolvable, intractable problems or challenges. Building a technology company in the middle of the sea has been no simple undertaking, but it's been my great pleasure to work with this inspiring team. It's personally gratifying to work with people I admire and respect in a place that I love called Hawaii.

I would also like to acknowledge many, many people that have mentored or supported me along the way, including too many to name, but to just mention a few: My lovely wife Jan Naoe Sullivan, who is my most trusted adviser, Mr. Barry Weinman, a seasoned venture capitalist and entrepreneur, Dr. John Craven, the quintessential renaissance man who introduced me to transdisciplinary thinking, Professor Bruce Liebert, my PhD advisor and mentor, Dr. Hans Krock, Senator Daniel Inouye, Dr. Alfred Yee, Mr. Dudley Pratt, Mr. John Dean, Mr. Roy Takeyama — my father in law, and other mentors, faculty and professors that have taken an interest in me.

Putting a book like this together was a long and arduous undertaking and would not have been possible without the support and mentorship from writing and publishing professionals, including Roger Jellinek, Tony Pisculli, as well as Michelle Katchuck.

ABOUT THE AUTHOR

An entrepreneur, scientist, businessman, technologist, engineer, founder, public speaker and futurist, Patrick Sullivan founded Oceanit in 1985. Securing more than $500M for its programs, Oceanit's engineers, scientists and professionals have delivered innovative solutions in the fields of aerospace, energy, bio-photonics, artificial intelligence, therapeutics, nanotechnology and optics, as well as technology for missile defense, advanced manufacturing, space debris, and environmental sustainability.

Dr. Sullivan, named Hawaii Business News' *"2019 Titan of Technology Leader,"* and Hawaii Business Magazine's *"2016 CEO of the Year,"* developed Oceanit's "Mind-to-Market" business model that integrates "Design Thinking" for human-centered innovation with "Intellectual Anarchy™," the art of disruptive innovation. He also has launched and enabled numerous private equity financed start-up and spin-out companies.

Dr. Sullivan has extensive experience on boards and commissions, including the Secretary of the Navy's Ocean Research Advisory Panel, the Rehabilitation Hospital of the Pacific and The Pacific International Space Center for Exploration Systems (PISCES). He serves on Rice University's Material Science and Nano Engineering (MSNE) Advancement Committee and the Dean's Engineering Council for the College of Engineering for the University of Hawaii. He has also successfully developed and launched national research programs with DARPA.

Dr. Sullivan lives in Hawaii and enjoys surfing and all water sports. He and his family are avid outdoorspeople and enjoy everything from snow skiing to rock climbing, as well as traveling to remote, uncharted places around the world with little more than a backpack.

www.ingramcontent.com/pod-product-compliance
Lightning Source LLC
Chambersburg PA
CBHW071330210326
41597CB00015B/1397